Optical
Engineering
Fundamentals

Optical and Electro-Optical Engineering Series

Robert E. Fischer and Warren J. Smith, Series Editors

Published

Hecht
THE LASER GUIDEBOOK

Manning
STOCHASTIC ELECTROMAGNETIC IMAGE PROPAGATION

Nishihara, Haruna, Suhara
OPTICAL INTEGRATED CIRCUITS

Rancourt
OPTICAL THIN FILMS USERS' HANDBOOK

Sibley
OPTICAL COMMUNICATIONS

Smith
MODERN OPTICAL ENGINEERING

Smith
MODERN LENS DESIGN

Waynant, Ediger
ELECTRO-OPTICS HANDBOOK

Wyatt
ELECTRO-OPTICAL SYSTEM DESIGN

Other Published Books of Interest

Optical Society of America
HANDBOOK OF OPTICS, VOLUMES I, II

Keiser
OPTICAL FIBER COMMUNICATIONS

Macleod
THIN FILM OPTICAL FILTERS

Syms, Cozens
OPTICAL GUIDED WAVES AND DEVICES

To order or to receive additonal information on these or any other McGraw-Hill titles, please call 1-800-822-8158 in the United States. In other countries, please contact your local McGraw-Hill office.

BC14BCZ

Optical Engineering Fundamentals

Bruce H. Walker

Walker Associates, President

McGraw-Hill, Inc.

New York   San Francisco   Washington, D.C.   Auckland   Bogotá
Caracas   Lisbon   London   Madrid   Mexico City   Milan
Montreal   New Delhi   San Juan   Singapore
Sydney   Tokyo   Toronto

Library of Congress Cataloging-in-Publication Data

Walker, Bruce H.
    Optical engineering fundamentals / Bruce H. Walker.
        p.   cm. — (Optical and electro-optical engineering series)
    Includes index.
    ISBN 0-07-067930-4
    1. Optics.   I. Title.   II. Series.
TA1520.W35   1994
621.36—dc20   94-21244
                                                          CIP

1 2 3 4 5 6 7 8 9 0   DOC/DOC   9 0 9 8 7 6 5 4

ISBN 0-07-067930-4

The sponsoring editor for this book was Steve Chapman, the editing supervisors were Kimberly A. Goff and Joseph Bertuna, and the production supervisor was Pamela A. Pelton. This book was set in Palatino. It was composed by McGraw-Hill's Professional Book Group composition unit.

Printed and bound by R. R. Donnelley & Sons Company.

Contents

Foreword

In *Optical Engineering Fundamentals* Bruce Walker has undertaken, and successfully completed, an immensely difficult task, that of producing a palatable, yet correct, introduction to the fundamentals of optical engineering. He has drawn on his years of practical, real-world experience as a lens designer and optical engineer to select the appropriate topics for inclusion in this work, and has utilized to great advantage his unquestioned flair for simple, easy-to-understand exposition of relatively complex subjects. The result is a book which will serve as an excellent introduction—a "primer" in Bruce's own words—to the field. It provides not only a solid grounding in optical engineering which serves as a bridge to, and a basis for, understanding more advanced works; it is also a useful, usable manual for those who want to begin the practice of optical engineering.

Optical Engineering Fundamentals is a welcome addition to our Optical and Electro-Optical Engineering Series.

Warren J. Smith

Preface

The concept of this book, along with many of the ideas and much of the material contained here, has evolved over the last 25 years. During that time, while working in the field of optical engineering and lens design, I have frequently been called on to describe and explain certain optical theories or phenomenon to coworkers, or to readers of trade publications. In responding, my goal has always been to reduce these explanations to the most basic technical level, one that has been easily comprehensible. The value of this book, then, is not in bringing forth new information never previously available to the reader. Rather, it is a carefully thought-out selection of that material which it is felt will be of maximum interest and value to the reader. That material will be presented in a form that can be easily understood, in the absence of complex theories of mathematics and physics.

The field of optical engineering and the subject of optics in general, are not merely interesting, but often quite fascinating. In order to be a *good* optical engineer, one must have a sincere interest and curiosity about the subject. This must then be supplemented by a fundamental knowledge of just a few very basic principles that will allow that curiosity to be satisfied. This book is designed to assist the student, or worker who is interested and involved in the field of optics, to obtain a better understanding of those basic principles and to prepare the reader for the more complex topics that will be encountered and dealt with using more advanced, specialized textbooks and reference material.

Thanks to the extraordinary nature of the human visual system and the many wonders of the world in which we live, hardly a day passes that we are not exposed to the science of optics, albeit often without a full understanding of that which we are experiencing. Consider modern technology such as home television systems that include projection TVs, laser disks, and compact camcorders. In the case of home audio systems, consider the compact laser disk revolution that, in just a few short years, has made the traditional phonograph record essentially obsolete. Consider also the optical technology that has been demonstrated by the space and defense industries in recent years. We have been privileged to view pictures from space and from distant planets. Military conflicts have been decided by the use of "smart" bombs, laser sights and head-up displays...all sophisticated products of today's optical engineer.

These are wonders of our own making...we need only observe a colorful sunset, or view a rainbow, to witness some of the many optical wonders of nature. Of course we would not observe this or anything else were it not for the most amazing of nature's optical systems: the human eye and its associated physiological components. It is my sincere hope that this book will, in some small way, make it possible for the readers to understand and appreciate many of these things, so that they might feel more a part of all that goes on around them, especially as it relates to the science of optics and the field of optical engineering.

If you work in the field of optics, I'm certain you will find that a better understanding of the subject will serve to improve your basic skills. In addition, I believe it will also enhance your enjoyment of not just your work, but hopefully of life in general. Equally important, it is this understanding that makes it possible for us to share these experiences with others.

One goal of this book is to demonstrate that there are many aspects of the science of optics and the field of optical engineering that are not that difficult to understand. Until now, the majority of texts and other publications in this field have assumed a certain *level of understanding*, and proceeded from that point. In some 25 years of work as an optical engineer, which has included the preparation and presentation of numerous technical papers and articles for a variety of publications, I have encountered a very real need, and a sincere desire on the part of many, to obtain that information required in order to reach that assumed level of understanding. This book is presented in the belief that it will serve to meet that need and to quench that desire.

Bruce H. Walker

Acknowledgments

One of the many advantages of a career in the field of optical engineering is the assurance that one will encounter a wide variety of very talented and interesting people in the course of one's work. As a young, fledgling engineer at General Electric in the early 1960s, my interest in optics and lens design was sparked and nurtured by three such individuals. For their early and sustained inspiration and support, I would like to thank Dr. Jack Mauro, Mr. Bob Sparling, and Mr. Don Kienholz.

Optical
Engineering
Fundamentals

1
Introduction

Optical Engineering Fundamentals is intended to be a "bridge" book...a book that will carry the reader who is interested in the field of optical engineering from a position of curiosity, confusion, and apprehension to one of understanding, comfort, and appreciation. While this book alone will not make you an optical engineer, it is hoped that it will enable you to communicate with and work closely and effectively with the optical engineer. Should it be your goal to eventually work in the field of optical engineering, this book will provide major assistance in two ways. First, it will provide a basic understanding and appreciation of many fundamental optical principles, thus preparing you to get the most out of the many fine advanced optics texts and courses of study that are available. Second, as you work into the field, this book will serve as a convenient source of valuable reference material.

How many of us have said to ourselves (or heard it said), "I don't know anything about optics, but..." and then the speaker has proceeded to conclusively prove the point. It is the author's fondest hope that this book will render that phrase obsolete for its readers. Frequently, an *optics* subject is brought up in a discussion, couched in terms that are esoteric beyond the understanding of the layperson or novice. This book will provide the reader with a level of knowledge that will assure a degree of comfort in discussing the basics of such topics and exploring their more advanced aspects, as required.

For many years the job descriptions for the optical engineer and the lens designer were considered to be quite separate and distinct. Because the science of lens design was so unique and because the calculations involved were so time-consuming, the lens designer rarely had the time or inclination to venture into the more general fields that were the responsibility of the optical engineer. Likewise, the successful optical engineer has traditionally been more than willing to leave the tedious task of lens design to the specialist. Recent developments in modern technology have led to a significant realignment in these areas.

With today's modern personal computer systems and easy (read "fun")-to-use software packages, dedicated to lens design and optical analysis, it behooves today's optical engineer to develop a certain level of competence in this area. While the intricacies of starting lens type selection and detailed lens optimization are still best left to the lens designer, the ability to utilize a modern optics program for performing basic optical system analysis will make one a much more effective optical engineer. Likewise, the speed with which lens design tasks can now be accomplished allows lens designers to diversify their activities into some of the more general areas of optical engineering. The two classifications that once were quite distinct and clearly defined are gradually blending together into a single position, often labeled *optical designer*. This is good for the optics industry, in that there has always been a shortage of qualified people in both areas. More important, it is very good for individuals who select this as a career path, in that it broadens the scope of their work, making it more interesting, challenging, and rewarding.

It is believed that anyone working in or dealing closely with the optics industry will find this book to be both interesting and useful. A first reading will create a sense of comfort with the subject matter not obtainable from previously available material. As a teaching tool this book will prove most valuable in exposing the student to the intriguing science of optics and optical engineering, without the complications of advanced physics and mathematical theory. Subsequently, the formulas, illustrations, and tables of information contained within the book will serve as valuable reference material for all working within, or in close proximity to, the field of optical engineering.

The book begins with an historical review, covering some of the more interesting people and events that have been involved in the study and application of the science of optics over the years. While not intended to be comprehensive, this historical information will be of interest in terms of the relationship between modern optical engineering and the events and people of the past.

Similarly, an early chapter in the book deals with the age-old question, "What is light?" While a definitive answer is not forthcoming, the presentation should result in a certain *level of comfort* with the topic and will allow the reader to more confidently and intelligently join with the likes of Aristotle, Ptolemy, and Galileo in pondering the ultimate answer to this most basic question. The relationship between light waves and light rays is then presented. The laws of reflection and refraction and their relationship to basic optical components are introduced, as is the subject of dispersion. The topic of diffraction is also touched on, to the extent that it affects the performance and image quality of many typical optical systems.

The topic of thin-lens theory is developed and discussed, with several examples of how it can be applied to real-world problem solving. This is followed by an introduction to an all-new computer software package, designed to aid the optical engineer in understanding and evaluating the optical system. Several examples of thin-lens theory application are included. For the transition from a preliminary design to a final configuration, a basic understanding of the primary aberrations of a lens system is most helpful. Coverage of this topic includes a basic description of these aberrations with a discussion of their origins and impact on performance and image quality of the lens system.

Having established this foundation, the book then devotes several chapters to a discussion of the design, function, and application of basic optical instruments. Details of a variety of typical optical components are discussed, including the manufacturing processes employed and the determination of specifications and tolerances. The characteristics, behavior, and relative strengths and weaknesses of optical materials are discussed. Included among these are optical glass, metal optics, coatings, and special materials for the ultraviolet and infrared portions of the spectrum. The special considerations of optical engineering as it relates to systems used *visually* are covered, including a comprehensive review of the general characteristics, configuration, and function of the human eye. A separate chapter is dedicated to modern lens design methods, including a brief historical review of computing techniques and the modern computer as it relates to the history of lens design. Finally, several of the more interesting optical devices and phenomena encountered in day-to-day life are covered.

If the reader comes away with one predominant sense after reading this book, I hope it will be an awareness of the many ways in which optics and optical engineering are interrelated with our everyday lives. Every book has its purpose, in this case that purpose is to present some of the more unique, interesting, and useful aspects of optics and optical engineering, in a format that will be enjoyable, useful, effective, and entertaining...rather than intimidating.

2

Historical Review

2.1 Definition of Optical Engineering

By definition, *optics* is the scientific study of light and vision, chiefly of the generation, propagation, manipulation, and detection of electromagnetic radiation having wavelengths greater than x rays and shorter than microwaves. The term *optical* applies to anything of or pertaining to optics. By further definition, then, *optical engineering* would be the application of scientific optical principles to practical ends, such as the design, construction, and operation of useful optical instruments, equipment, and systems. This chapter will deal with some of the people, places, and events that, over the years, have contributed significantly to the history of optical engineering.

2.2 Ancient History

The earliest indications of some knowledge and application of optical principles appeared nearly 4000 years ago when two unrelated massive structures, Stonehenge and the Pyramid of Cheops, were constructed. In both cases the orientation of these structures is found to be tied closely to the relationship between the earth and the sun. Familiarity with the cyclical relationship of the seasons, along with the knowledge that light travels in straight lines, would have been required for the resulting arrangements to occur. The precise orientation of these structures relative to the compass and the calendar may be interpreted as an early demonstration of basic optical engineering principles.

Early historical records of Plato and Aristotle (c. 350 B.C.) reveal one of the first debates over the exact nature of light. While Plato taught that vision was accomplished by the expulsion of *ocular beams* from the eyes, his student Aristotle rejected that theory, arguing (more correctly) that vision arises when particles emitted from the object enter the pupil of the eye.

While Euclid (c. 300 B.C.) is best known for his study of and writings dealing with geometry, he contributed a great deal to the science of optics through his book titled *Optics*. While erroneously accepting Plato's theory of ocular beams, Euclid did, among other things, correctly describe the formation of images by spherical and parabolic mirrors.

Knowledge of geometry and knowledge of optics were elegantly combined in an experiment intended to determine the circumference of the earth by a Greek scholar named Eratosthenes, while he was living in Egypt about 230 B.C. He noted that on a given day and time each year, the sunlight fell directly down a well in the village of Syene. Knowing the distance from Syene to Alexandria, some 800 km to the north, he erected a vertical pole there and carefully measured the length of the pole's shadow at that same precise time. Assuming that extensions of the well at Syene and the pole at Alexandria would intersect at the earths center (see Fig. 2-1), Eratosthenes was able to conclude that the earth's circumference was 50 times the distance from the well to the pole, or 40,000 km. This value is remarkably close to today's established value of 40,074 km.

Ptolemy (c. 150 A.D.) was another great scientist of that age. In his book *Optics*, Ptolemy carries on the ocular beam theory, while also describing correctly the law of reflection (the incident ray, the reflected ray, and the normal to the reflecting surface lie in a common plane, with the angle of incidence being equal to the angle of reflection). Ptolemy also studied refraction, and published a similar theory, but failed to establish the exact relationship between the angle of incidence and the angle of refraction.

2.3 Medieval Optics

Alhazen (c. 1000) was a preeminent Arabian scholar who examined and explained the anatomy and function of the eye. It is thought that his work may have inspired the invention of spectacles. Alhazen is

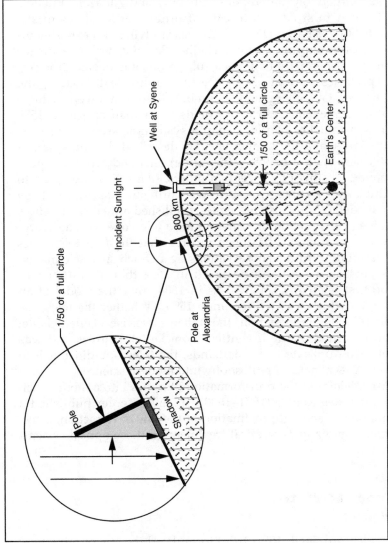

Figure 2-1. Around 230 B.C., knowledge of geometry and knowledge of optics were cleverly combined by Eratosthenes to estimate the size of the earth. Knowing the distance from Syene to Alexandria, and the relationship of the pole to its shadow when the sun was directly overhead at the well, he concluded that the earth's circumference was 50 times the distance from well to pole.

also credited by some with the invention of the camera obscura. There is a distinct functional similarity between the eye and the camera obscura (Fig. 2-2); this fact may have been instrumental in the work done by Alhazen.

During this same time period there was a revival of interest in optics in the west. Vitello of Poland wrote a treatise on optics around 1270. While based largely on the works of Ptolemy and Alhazen, Vitello's work did much to stir the interests of other medieval scientists. Around 1260, Roger Bacon (1214–1292) was studying and experimenting with lenses to improve vision. Whether Bacon deserves credit for the invention of spectacles remains a subject of some debate. It is certain that spectacles were in use by the time of his death. Led by the likes of Chaucer, Gutenberg, and Columbus, these years were witness to significant achievement in many fields other than optics. In 1550 Francesco Maurolico of Naples began a systematic study of prisms, spherical mirrors, and the human eye. He described the correction of nearsightedness with a concave (negative) lens and farsightedness with a convex (positive) lens. As had others, Maurolico attempted to derive the law of refraction but was not successful. In 1589 Giambattista della Porto, also of Naples, published a treatise on lenses that contained construction information for a telescope with a diverging (negative) eye lens. It has been written that the first telescopes were the result of fortuitous observations by a Dutch spectacle maker. Johannes Janssen of Holland declared in 1634 that his father had "made the first telescope amongst us in 1604, after the model of an Italian one, on which was written anno 1590." Whether the telescope was a Dutch or Italian invention, the Italian physicist, Galileo Galilei (1564–1642) learned of the invention from Dutch sources and was quickly able to duplicate it. In his hands, the telescope did much to transform medieval natural philosophy into modern science.

Also instrumental in the transformation of optics into a modern science was Johannes Kepler (1571–1630). In 1604, Kepler published a work containing a good approximation of the law of refraction, a section on vision, and a mathematical treatment on those optical systems then known.

2.4 From 1620 to the 1900s

In 1621, Willebrord Snell (1591–1626) formulated the law of refraction. Independently, and at about the same time, the same law was formu-

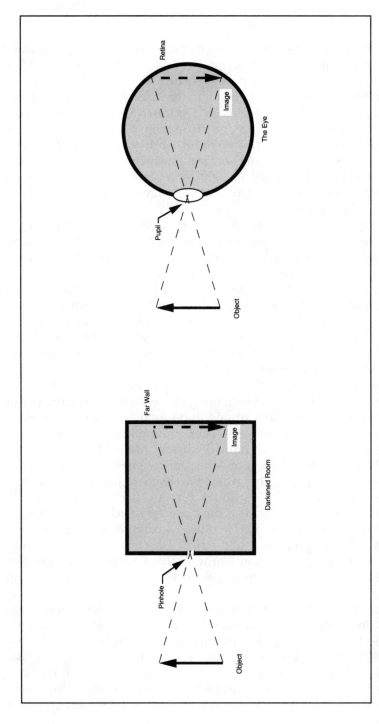

Figure 2-2. The camera obscura (left) and the eye (right) were both studied by the Arabian scholar Alhazen, c. 1000. The optical similarity of the two is shown in this illustration.

lated by Rene Descartes and James Gregory. The honor of precedence was given to Snell and, as a result, the law of refraction carries his name today. Later in the seventeenth century a number of experiments were conducted which added greatly to the understanding of light and its behavior. Several experiments by Francesco Grimaldi (1618–1663) demonstrated the phenomenon of diffraction, although many years would pass before the significance of his work was completely understood. Sir Issac Newton (1642–1726) conducted experiments that demonstrated the dispersion of light by a prism. Newton's book on optics was first published in 1704 and would remain an influential reference for more than a century.

A large part of the history of optics from 1650 to modern times is represented by the independent development of two optical instruments and a related process: the telescope, the microscope, and photography. It was in 1668 that, while a 26-year-old graduate student at Cambridge, Isaac Newton wished to build a telescope for his own use. After carefully considering the potential problems of using lenses to build a refracting telescope, Newton opted to build a reflecting telescope with a metal mirror for an objective. While only 1.3 in in diameter with a 6-in focal length, this successful telescope represented a significant accomplishment for the time.

In these years, the use of large lenses for a telescope objective was frustrated by the lack of understanding of the dispersive properties of glass, and also by the lack of adequate quality in available glass blanks. It would be some 60 years later, around the year 1730, that an obscure 25-year-old lawyer named Chester Moor Hall (1704–1771) solved the problem of dispersion when he designed and had built the first achromatic doublet. This was a two-element lens that combined lead oxide glasses (flints) with the commonly available crowns. Still, the glass quality problem limited the diameter of achromatic lenses to just a few inches. In the early 1800s, a number of people were involved in developing new methods of glass manufacture to eliminate these problems. As a result, around that time Joseph Fraunhofer (1787–1826) was able to produce achromatic lenses as large as 8.5 in in diameter.

The mirrors used by Newton and others to build telescopes were polished into base metal substrates that would have to be frequently repolished in order to eliminate tarnishing. In 1850, J. von Liebig (1803–1873) invented a process for the chemical silvering of glass. After this, all mirrors were made of glass and simply stripped and resilvered when they became tarnished. It would be the 1930s before that process was replaced by the use of deposited aluminum, which essentially eliminated the tarnishing problem. Also in the 1930s, several revolutionary reflecting objective lens designs appeared. The Ritchey-Chretien and Schmidt configurations would greatly increase the limits of image quality and field of view for the reflecting telescope objective.

Paralleling the development of the telescope during these years was the development of the microscope. Devised originally by the Dutch father-and-son team of Hans and Zacharias Janssen in the late 1500s, the microscope would later be improved on by many others. The simplest version, containing just a single lens of very short focus, was described and used extensively by the Dutch biologist van Leeuwenhoek (1632–1723). An early form of the improved compound version of the microscope appears in the publication *Micrographica*, by Robert Hooke (1635–1703) in 1665. It was in the late 1700s that achromatic microscopes were first produced by Dollond and others. The problem of combining several corrected achromatic doublets to yield increased magnification was solved in 1830 by an amateur microscopist named Joseph Jackson Lister (1786–1869), father of the famous surgeon Lord Lister. Around 1850, the Italian optician G. B. Amici (1786–1863) added a planoconvex lens to the front of the microscope objective, permitting even greater magnification. In 1879, Ernst Abbe (1840–1905) established the relationship between numerical aperture and resolving power in a microscope objective. Abbe also introduced the concept of homogeneous immersion and later, the use of crystalline fluorite for the elimination of secondary color in microscope objectives. In the early 1900s there was a flurry of activity involving the use of ultraviolet illumination in microscopy to achieve the improved resolution that occurred as a result of the reduced wavelength. In 1928 this all became academic with the introduction of the electron microscope, which brought with it a 100 × (hundredfold) increase in resolving power.

The process of photography was a dream of many during these years, as the images of the camera obscura were observed but could not be recorded. Finally, in 1839, the French artist Daguerre (1789–1851) announced his system of photography, using an iodized polished silver plate which was developed with mercury vapor. For a lens with a wide flat field, Daguerre turned to his friend Charles Chevalier (1804–1859), who suggested the use of a reversed telescope objective with an aperture stop in front that would limit the lens speed to f15. The need for increased lens speed soon became apparent and in less than a year the 32-year-old mathematics professor at the University of Vienna, J. M. Petzval (1807–1891), designed his famous portrait lens with a speed of f3.5. This new design resulted in an increase in image brightness of nearly 20 times over that of the lens by Chevalier. From that time, and continuing today, the design of lenses has been greatly influenced by the needs and desires of the photographer. The wide-angle concentric design by Schroeder at Ross in 1887 and the Anastigmat by Paul Rudolph (1858–1935) at Zeiss in 1900 were monumental lens design achievements. Since that time, while the achievements in photographic lens design have been great in number, they have been more incremental in nature.

By 1781, Sir William Herschel (1738–1822) had utilized modern optical technology to discover the planet Uranus, nearly doubling the extent of the known solar system. While conducting further studies involving the sun, Herschel discovered thermal activity in dispersed sunlight beyond the red end of the visible spectrum. Following much debate, Thomas Young (1773–1829) was responsible for experiments that confirmed the existence of infrared energy. Young's work went a long way toward explaining the wave nature of light, including the diffraction effects that had been observed by Grimaldi some 150 years earlier. In addition, as a physician of some note, Young did considerable research and reporting on the mechanism of the eye.

2.5 The Speed of Light

Over the years, the history of optics has been tied inexorably to the quest to determine the velocity of the propagation of light in various media. Initially, it was thought that light traveled with infinite speed. Earliest experiments indicating otherwise were related to astronomical observations. In 1675, the Danish astronomer Olaf Roemer (1644–1710) made the first scientific determination of the speed of light on the basis of his observations of the eclipses of the innermost moon of Jupiter. Roemer noted a significant difference in the times at which these eclipses occurred, depending on the relative position of the earth in its orbit around the sun. Essentially, when the earth was nearest to Jupiter the eclipses would occur several minutes ahead of the predicted time and when the earth was farthest from Jupiter, that occurrence would be several minutes later than had been predicted. While there is no record that Roemer actually did the final calculation, his data would lead to the conclusion that light travels at a speed of about 200,000 km/s. Contemporaries of Roemer would modify his findings, including more accurate data on the earth's orbital radius, and come up with a value closer to 300,000 km/s.

Approximately 50 years later, in 1728, the noted British astronomer James Bradley (1693–1762) made an entirely different kind of astronomical observation, from which he was able to compute the velocity of light. This experiment involved the observation of a star with a telescope whose axis was set perpendicular to the plane of earth's rotation. It was found that in order to compensate for the speed of the incident light, the telescope axis would have to be tilted through a small angle in the direction in which the earth was traveling. The amount of telescope tilt required allowed Bradley to compute the velocity of light, which he found to be 300,000 km/s.

The first terrestrial measurement of the speed of light was done by the French scientist Fizeau in 1849. Fizeau's experiment is illustrated in Fig. 2-3. A light source was focused through a beamsplitter onto an image plane where a spinning toothed wheel was located. The light passing through that wheel was then projected to a mirror at a distance of several kilometers, where it was collected and then reflected back to its point of origin. The toothed wheel created a pulsed light source whose frequency was determined by the number of teeth on the wheel and the rate at which it was rotated. When the wheel's speed of rotation was set such that the light pulse returned and passed through the very next space in the wheel, Fizeau was able to compute the velocity of that light pulse. Limited by the precision of his measurements, Fizeau calculated the speed of light to be 315,000 km/s. Fizeau's experiment was later modified by the French physicist J. L. Foucalt (1819–1868), who replaced the toothed wheel with a rotating mirror. With this arrangement Foucalt determined the speed of light to be 298,000 km/s, much closer to today's accepted value. Foucalt was also able to insert a tube filled with water between the rotating mirror and the distant mirror and determine conclusively that the velocity of light was reduced when traveling through water. This conclusion went a long way toward dispelling the corpuscular theory, which demanded that the speed of light in water be greater than it was in air. The Foucalt method was further improved by many, with the most precise measurements made by the American physicist A. A. Michelson (1852–1931). The average value of a large number of measurements done by Michelson was 299,774 km/s. Many aspects of modern technology have been applied to the determination of the speed of light in recent years, yielding a currently accepted value of 299,793 km/s. Finally, it is interesting to note that electromagnetic theory allows the velocity of electromagnetic waves in free space to be predicted, with a resulting value of 299,979 km/s, which is within 0.1 percent of the most precise measured values.

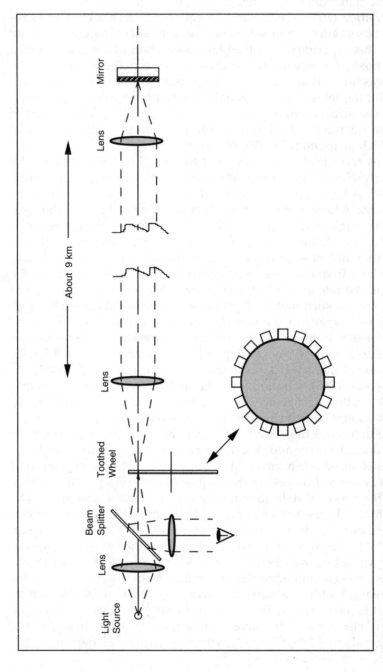

Figure 2-3. The first terrestrial measurement of the speed of light was done by Fizeau in 1849 when he projected a pulsed beam of light to a distant mirror. Knowing the number of teeth and speed of rotation of the toothed wheel and the distance to the mirror, he was able to calculate a value of 315,000 km/s.

2.6 Modern Optical Engineering

Optical engineering in the twentieth century has witnessed many sweeping technological advances. A review of developments relating to the photographic lens system will serve well to demonstrate one such area. At the turn of the century photography had been around for some 50 years, and only a few very basic lens types were available to the photographer. A better understanding of lens aberration theory and the mathematical methods of ray tracing would now permit the design of more sophisticated lens forms, such as the Cooke Triplet by H. Dennis Taylor (1862–1943) and the Zeiss Tessar lens by Paul Rudolph (1858–1935). Testimony to the significance of these designs is the fact that both remain in common use by photographers today, nearly 100 years later. Another monumental design was the Lee Opic lens, designed by H. W. Lee (1879–1976) in 1920. This was an unsymmetrical double-gauss lens, derivatives of which are supplied as the *standard* lens on many of today's 35-mm cameras. Figure 2-4 illustrates these three classic lens designs.

From around 1950, to and including today, the variable focal length or zoom lens has been through a most fruitful period of development. All the most significant technological breakthroughs in a number of fields have been applied to the production of today's high-performance, compact zoom lenses for television, motion picture, and still photography.

The development of these high-performance lenses would not have been possible without simultaneous advances in the field of optical glass manufacture. In Europe and Great Britain several precision glass manufacturers were in place at the turn of this century. Soveril, in France, had a history dating back to 1832, while the roots of the Chance-Pilkington glass company in England could be traced back to 1824. In Germany, the Schott Glass Works had begun its work in 1880. In the United States, Bausch and Lomb established a glass manufacturing capability in 1912. This proved most fortunate when in 1914 the start of World War I cut off all supplies of available optical glass from Europe. At Eastman Kodak, a program was begun in 1937 that involved the manufacture of special optical glasses formulated specifically for the purpose of producing improved lens designs. Since the end of World War II, the Japanese have become major producers of high-quality optical glasses. This is one of the factors that has made possible their virtual domination of the 35-mm camera lens market. The two principal Japanese glass manufacturers today are OHARA and HOYA.

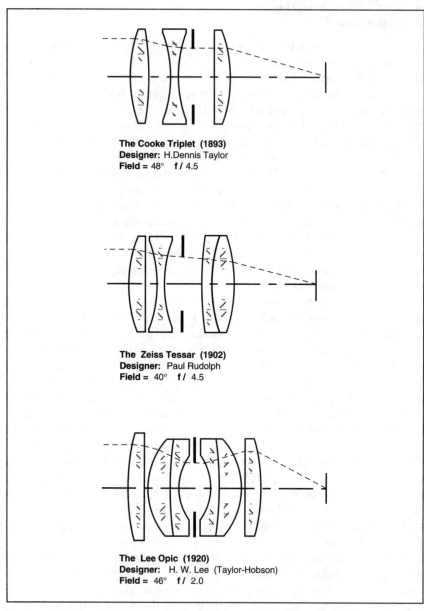

The Cooke Triplet (1893)
Designer: H.Dennis Taylor
Field = 48°   f / 4.5

The Zeiss Tessar (1902)
Designer: Paul Rudolph
Field = 40°   f / 4.5

The Lee Opic (1920)
Designer: H. W. Lee (Taylor-Hobson)
Field = 46°   f / 2.0

Figure 2-4. Three milestone designs in the history of photographic lens design. Each of these basic lens design forms remains in common use today.

Another area of optical technology that has evolved dramatically during this century is that of optical coatings to reduce reflection. Early lens designs involved just a few air to glass interfaces. As a result, the amount of incident light lost by surface reflections was not significant. In the case of a telescope objective in the form of a cemented doublet, only two surfaces were significant in terms of overall light transmission. As photographic lens designs became more complex, lenses such as the Cooke Triplet with six air–glass surfaces became common. In this type of lens, surface reflections would reduce the overall light transmission to about 70 percent. While the 30 percent light loss was important in its own right, another problem existed in that much of the reflected light would eventually find its way to the image plane in the form of stray light and ghost images, thus greatly reducing image contrast and clarity. Perhaps influenced by the fact that these reflections detracted from his otherwise outstanding lens design, H. Dennis Taylor (designer of the Cooke Triplet) was instrumental in early studies of antireflection coatings. Taylor observed in 1896 that certain lenses that had become *tarnished* by environmental exposure exhibited improved light transmission characteristics. Around this same time, while working at Ross, Ltd. in England, Frederick Kollmorgen (1871–1950) is said to have observed similar improved light transmission through portions of a magnifying lens that had been accidentally stained with spilled ink. Various methods of chemically coating optics were attempted by Taylor, Kollmorgen, and others with varying degrees of success due to the unpredictable nature of the process. It was in 1936 that A. Smakula (1900– ?) of Zeiss invented the process of coating lens surfaces in a vacuum with a thin evaporated layer of a low-index material such as calcium fluoride or magnesium fluoride. This same process was also suggested by John Strong of California at about the same point in time. Subsequently, as the evaporated coating process was perfected, it was found that multiple layers of certain materials would further reduce the amount of reflected light. With these modern antireflection coatings it has become possible today to produce a complex lens assembly with 20 air–glass interfaces that will demonstrate essentially the same light transmission characteristics as an air-spaced doublet without coatings.

To be discussed in greater detail in a later chapter, the process of lens design will be touched on here from an historic perspective. The situation regarding lens design in 1840 can be best illustrated by the story relating how Joseph Petzval was assigned the help of two artillery corporals and eight gunners who were "skilled in computing" to aid him in the calculations required to complete a lens design. At the end of a 6-month period this team had completed two basic lens designs; one of these was the Petzval portrait lens.

Little had changed by the beginning of this century. Those involved in lens design employed mathematical formulas for the computation of aberrations and the exact tracing of rays through lens surfaces. The accuracies required called for the use of seven-place logarithms and for calculations to be executed with extreme caution. Frequently, as was the case with Petzval, teams of several individuals were assigned to the designer to execute these tedious calculations. It was in the 1930s that the mechanical calculator was developed and this burden was just slightly reduced. Still, the design of a complex lens would involve the work of many people, working over periods of months and often years.

Finally, in the 1950s, the electronic computer appeared on the scene, and became available for the solution of lens design problems. In the period from 1960 to 1990 the progress in the area of computing hardware and software has been nothing short of remarkable. Today, a skilled lens designer, equipped with a personal computer and appropriate software, can duplicate the work done in 6 months by Petzval and his team of corporals and gunners, in a matter of minutes.

Any discussion of optical engineering in the twentieth century would not be complete without mention of the laser. This device, with its capability of producing pure monochromatic light and a perfectly collimated light bundle, has had an immeasurable impact on the field. Devices such as interferometers, laser range finders, holograms, laser disk players, and so many more, have become a most important part of optical engineering in the twentieth century.

2.7 A Case History: Optics in the United States

Examination of the historical time line in Fig. 2-5 shows that prior to the twentieth century there was little activity dealing with optics and optical engineering in the United States. Since the turn of the century an increasing amount of research and engineering has been done in this country. Following the history of one optical company in the United States during that time period will provide some interesting insight into this period of growth and development.

In 1905, Frederick Kollmorgen left his position of optical designer with Ross, Ltd. of London to come to the United States to work with Keuffel & Esser in Hoboken, New Jersey as a lens and instrument

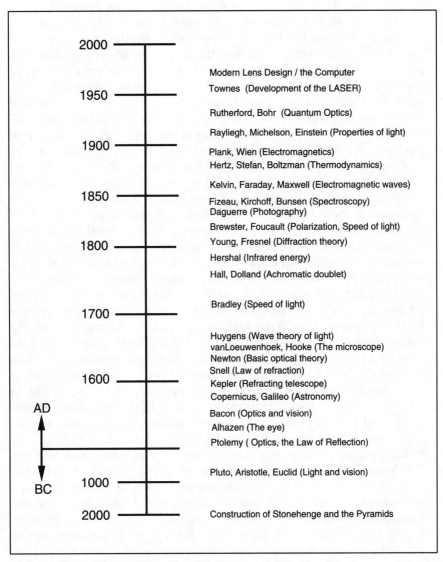

Figure 2-5. An historical time line of the history of optical engineering.

designer. Kollmorgen developed a keen interest in the subject of periscope design as it was applied to military systems and submarines. This led to several patents in this area being assigned to him in 1911. In 1914, with World War I eminent, the U.S. Navy contacted Keuffel & Esser in the hope that they might become the supplier of periscopes for the submarine fleet that was then under development. Keuffel & Esser declined the proposal, but Kollmorgen's interest in the project led him to seek an alternate approach. Kollmorgen had been dealing with a small optical company in New York City named Eastern Optical. He approached them with the proposition that they join forces to become the Navy's periscope supplier. The favorable business aspects of such a venture were immediately obvious, and the Kollmorgen Optical Company (KOC) was formed early in 1916. The wisdom of this judgment is validated by the fact that every submarine periscope deployed by the U.S. Navy since that day has been designed by Kollmorgen personnel and well over 95 percent of them have been manufactured by Kollmorgen. During World War I, the manufacture of periscopes and other military instruments resulted in an increase in the number of employees at KOC to nearly 100 and a move from Manhattan to a larger facility across the river in Brooklyn. With the end of the war, KOC branched into commercial products, which included production of a very successful line of motion picture projection lenses. Late in the 1920s the great depression began and the very survival of the Kollmorgen Optical Company was threatened. In fact, for a brief period in 1932 the company was completely closed. In the mid-1930s the government sponsored several programs dealing with the development of military instruments and KOC began to rebuild. In the late 1930s, they were involved in the design and manufacture of periscopes, driftmeters, bombsights, and navigational instruments. With the start of World War II in the early 1940s manufacturing activity grew at a frantic pace. Between 1940 and 1945 the number of employees grew from 50 to more than 600. At the point of peak production in 1944, submarine periscopes were being produced at a rate of one per day. In addition, hundreds of less complex optical components and instruments were being produced daily. In 1945, World War II ended and with it the demand for large quantities of military equipment. KOC again went through the "swords to plowshares" routine, again finding the motion picture projection lens to represent a valuable commercial product line. Other commercial products included riflescopes and monocular viewers. During the 1950s there was considerable activity in the development of new motion picture formats such as Cinemascope and Cinerama. KOC was involved in much of this development activity, which would eventually lead to KOC receiving several Academy Awards for technical excellence in the field. During the 1960s the cold war and the start of the space race led KOC into a number of interesting fields, most involving periscope designs of one form or another. They also developed a line of optical tooling instruments that were well received at the time. Meanwhile, the submarine periscope business continued to grow, spurred on by the Polaris and subsequent Fleet Ballistic Missile programs of the U.S. Navy. This periscope design work peaked in the late 1960s when KOC designed and produced a state-of-the-art periscope, designed specifically for photographic reconnaissance applications. It was this project that brought the optical engineers at KOC into the modern computer age. Prior to the late 1930s, all lens design calculations had been done by hand, under the close scrutiny of the company's founder, Frederick Kollmorgen. During World War II, the design staff was expanded to include several additional optical engineers, along with a team of assistants working with mechanical calculators. In the late 1950s the electronic computer came on the scene, enhancing the capabilities of the design staff. It was not until the late 1960s that the introduction of the mainframe computer, along with a modern lens design software package, would finally eliminate the need for a staff of calculating assistants to work with the optical engineers. In recent years KOC has grown and diversified, becoming a multinational corporation, involved not just in optics, but also a variety of other product lines. Meanwhile, it has retained one division that is descended directly from the original company. There, much of the work begun by Frederick Kollmorgen in 1916 is still carried on today. This includes design and production of submarine periscopes for the United States and several foreign navies, along with a variety of other electrooptical products intended primarily for military applications.

2.8 A Recent Incident: The Hubble Telescope

Until very recently, a declaration that one earned a living as an optical engineer would nearly always precipitate a discussion of the fit and function of eyeglasses. Early in the 1990s, thanks to the Hubble Space Telescope program, all of that changed abruptly. Suddenly the eyeglass discussion has been replaced by one dealing with "What happened to the Hubble" and "Did you have anything to do with it?" A review of what has come to be known by some as the "Hubble trouble" will serve well to illustrate the function of the modern-day optical engineer and, unfortunately, several of the things that can go quite drastically wrong with a complex optical program.

Since the earliest times, the quality of images viewed by astronomers has been limited by the fact that the celestial objects had to be observed through the earth's atmosphere. Inconsistencies in the density of the atmosphere, combined with impurities and local stray light, tend to reduce the image quality of even the finest telescopes. This has resulted in the location of most major observatories away from city lights and on the highest available mountaintops. This concept would be epitomized by the placing of a large telescope in orbit around the earth, above its atmosphere. Following several preliminary efforts which realized varying degrees of success, the Hubble Space Telescope (HST) program finally got under way in 1977. The space shuttle would ultimately deliver into orbit, a Cassagrain telescope with an aperture of 2.4 m. After numerous delays, including the Challenger disaster in 1986, the HST was finally launched in April 1990. In June it was found that the telescope was not performing as anticipated. The image quality was not border-line bad...it was terrible. It was concluded by examination of the images being produced that one of the two mirrors in the telescope was faulty in its basic shape. An examination of test data and procedures revealed that it was indeed the primary mirror that had been produced using test apparatus that was seriously in error.

The primary mirror of the HST is a concave hyperbola in cross section. In order to test this shape, a precision null lens assembly was designed and produced that would generate a wavefront which precisely matched that hyperbolic shape. During the manufacture and assembly of that null lens a very fundamental mistake was made which resulted in an error in the location of one of its optical elements. The resulting null lens assembly would still function in what was apparently normal fashion. The problem was that the wavefront being generated by the null lens assembly was not per the design, and the optician charged with the primary mirror machining would shape the mirror surface such that it precisely matched that erroneous wavefront. In hindsight there are any number of steps that could (or should) have been taken that would have uncovered this problem. Indeed several incidents reported to have occurred during manufacture were clear indicators that there was a problem. Reportedly, schedule and budget considerations were such as to override these suspicions. A second, backup primary mirror was being manufactured simultaneously by a second manufacturer, using independent methods and test equipment. An interchange of the two mirrors would have revealed the problem. Finally, a test might have been performed following the assembly of the primary and secondary mirrors to confirm their combined image quality. This was not done, again apparently for reasons of budget and schedule. The final result was a telescope placed in orbit, which did not come close to meeting the performance levels that were anticipated and that it should have been capable of. One very important lesson to be derived from all of this is that care must be exercised at all levels of design, manufacture, and test in order to assure a successful outcome. In this case, a relatively obscure lens element was placed 1.3 mm away from its proper axial location. This in a system with major dimensions that are many meters. Everything else was done correctly (technically if not politically), the mirror surface was polished to an accuracy of 0.00001 mm, but using the wrong gauge. As a result, the image of a distant star is more than five times larger than had been expected. This level of error essentially offsets all the advantages that can be realized by placing the telescope in orbit above the earth's atmosphere. It is typically the job of the optical engineer to assure that disasters of this type and magnitude do not occur. Most familiar with the Hubble story agree that, had those optical engineering personnel involved in this program been allowed to execute additional tests indicated to be required, then the ultimate problem could have been avoided.

It is gratifying to report that, as this book goes to print, the Hubble team at NASA has successfully implemented a correction to the optics of the orbiting space telescope, and it appears that the project will now be able to accomplish all the original goals established for the program.

2.9 Summary

These paragraphs represent just a small part of all that has transpired over the years relating to the history of optical engineering. It is obvious that the field has been in existence in some form for a long time and that many new and exciting developments have been witnessed in recent years. Today we are in the midst of an explosion of technology. Whether we find ourselves working actively in the field, preparing ourselves to become workers, or merely persons living among and benefiting from the fruits of this science, there is much going on here to be understood and appreciated. With that in mind, let's get on with our study of optics and optical engineering.

Bibliography

Much of the historical information contained in this chapter has been derived from the following reference material:

D. J. Lovell, *Practical Optics,* an unpublished manuscript written in the 1960s.

R. Kingslake, *A History of the Photographic Lens,* Academic Press, 1989.

R. Kingslake, *Some Impasses in Applied Optics,* JOSA, Feb. 1974.

Jurgen R. Meyer-Arendt, *Introduction to Classical and Modern Optics,* Prentice-Hall, 1972.

3

Basic Concepts of Light

3.1 Light: An Elusive Subject

In order to begin to understand the field of optical engineering and the function of the optical engineer, it is important that we first develop some sense of what light is and how it behaves. It has been the experience of many, including this author, that this understanding must be neither comprehensive nor precisely correct to be valuable to optical engineers in the execution of their day-to-day responsibilities. After all, as we have seen in the previous chapter, many of the great minds throughout history have made very significant contributions to the science of optics while maintaining incomplete or, in some cases, completely erroneous ideas regarding the nature and behavior of light.

3.2 Understanding Light

First, we can safely and correctly state that *light is energy*. We know that light is electromagnetic energy and that, in terms of wavelength, it represents just a very small part of a broad electromagnetic spectrum (see Fig. 3-1). It will be helpful to pursue our understanding of the term *wavelength* a bit further at this point. In one general and very useful definition, light is that portion of the electromagnetic spectrum that the human visual system is capable of detecting. That is to say, it is the electromagnetic energy that we can see. Under normal daytime condi-

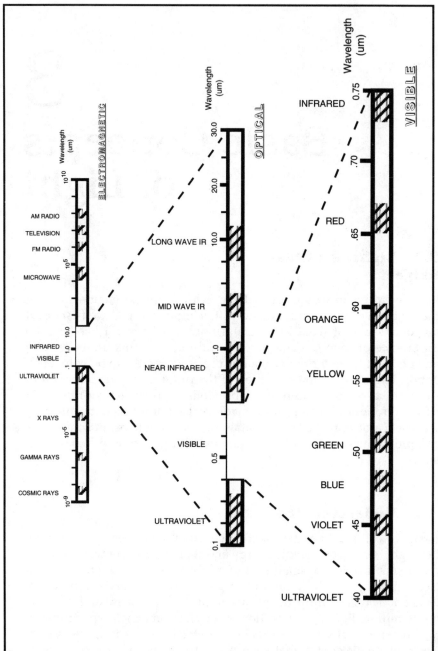

Figure 3-1. Wavelengths of the electromagnetic, optical, and visible spectra.

tions, the human visual system has a maximum sensitivity to light with a wavelength of 0.56 micrometers (μm). Let us convert this number (0.56 μm) to a more meaningful and comprehensible form. A meter represents a length of about 40 in, or a little more than one yard. A millimeter is $\frac{1}{1000}$th of a meter, the thickness of about 15 pages in this book. A micrometer (μm) is $\frac{1}{1000}$th of a millimeter and the peak wavelength for visible light is about half of that. Therefore, one wavelength of visible light is approximately equal to $\frac{1}{150}$th the thickness of a single page in this book.

As we continue the study of optics we will find that the dimensions we encounter are most often represented by either very large or, as is the case for wavelength, very small numbers. In order to deal with these numbers, it is helpful to understand and to utilize the system of scientific notation, which incorporates exponents. In its simplest form this method presents us with a number between 1 and 10, followed by a notation telling us how many places the decimal point must be moved to either the left (−) or right (+). In this case we might say the peak visible wavelength is equal to 5.5×10^{-4} (0.00055) mm. For those readers who are not yet thinking and working in metric terms, this would convert to 2.2×10^{-5} in (0.000022 in). As a first of many words of advice to the reader and would-be optical engineer, do make the conversion to the metric system: it will be most helpful and will enhance your effectiveness in nearly all areas of optical engineering.

As the wavelength of the energy collected by the eye increases or decreases, the color of that light as perceived by the eye will change. At the peak wavelength of 0.56 μm, the light is seen as yellow. When the wavelength decreases to about 0.50 μm the light appears green, while at 0.48 μm the color we see is blue. Moving in the other direction from the peak wavelength, at a wavelength of 0.60 μm, the light appears to be orange, and then, as the wavelength reaches 0.65 μm, we see the light as red. This is what we refer to as the *visible spectrum*, ranging from violet (0.45 μm) to red (0.70 μm), with a peak sensitivity to the color yellow, at a wavelength of 0.56 μm. The visible spectrum and its relationship to the electromagnetic spectrum are shown in Fig. 3-1. Figure 3-2 illustrates the relative sensitivity of the eye as a function of wavelength, or color.

3.3 Velocity, Wavelength, and Frequency

This wavelength concept may not be fully understood without some additional considerations. The topics of velocity and frequency, and

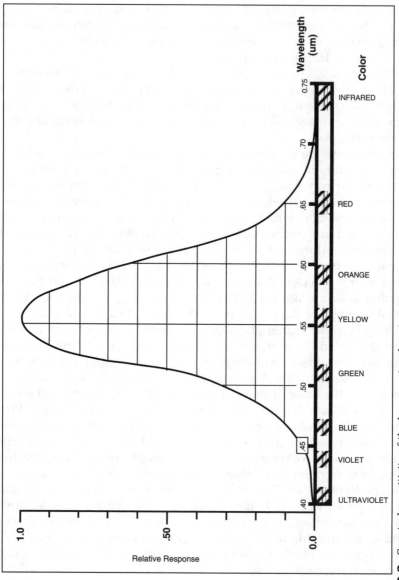

Figure 3-2. Spectral sensitivity of the human visual system.

their relationship to wavelength, must be introduced and discussed at this point. During our historical review of optics it was noted that, after many years of experimentation and several false starts, the velocity at which light travels in a vacuum was very accurately determined to be 299,793 km/s. For purposes of this discussion, a rounded-off value of 300,000 km/s (3×10^8 m/s) will be used to represent the speed of light in air. The enormity of this speed is best visualized when we understand that, traveling at this speed, we would be able to travel around the world seven and one half times in a single second.

Another example that will be helpful in illustrating the speed involved deals with recent experiments conducted to measure the distance from the earth to the surface of the moon. After the astronauts had placed a mirror assembly on the surface of the moon, it was possible to project a pulse of laser energy from here on earth, to that mirror and to detect its reflection when it had returned to its point of origin. Knowing the time involved for the round-trip, and the speed at which light travels, one could then calculate the distance from the source to the mirrors, i.e., earth to moon. Quite incredibly, the time required for the light to travel from the earth to the moon and return was only 2.5 s...that is fast!

In order to establish a relationship between the wavelength and the velocity of light that can be more easily understood and applied to our understanding of optics, the hypothetical example illustrated in Fig. 3-3 will be helpful. Consider a typical light source, such as the heated filament in an incandescent lamp. In order to further develop a sense for how light behaves, imagine this source to be pulsating, emitting light in a continuous wave within which the energy level is constantly and rapidly increasing to a maximum and then decreasing to a minimum. We have established that the light travels away from the filament at a speed of about 300,000 km/s. If a hypothetical energy-level detector were placed as shown, at a fixed distance from the source, it would register those maximum and minimum energy values as the wave impinged on the detector. By definition, the wavelength of this energy is the distance that the wave travels in the time that it takes the detector to record two consecutive maximum readings.

A heated filament is known to emit energy over a broad spectrum which includes the visible spectrum and part of the infrared band. To simplify our example, assume that a green filter is placed between the source and the detector as is shown in Fig. 3-3. Knowing the wavelength of the transmitted light to be 0.5 μm (5×10^{-7} m/cycle), and the speed at which this light is traveling (3×10^8 m/s), we can compute the frequency at which the detector will register maximum readings using the following formula:

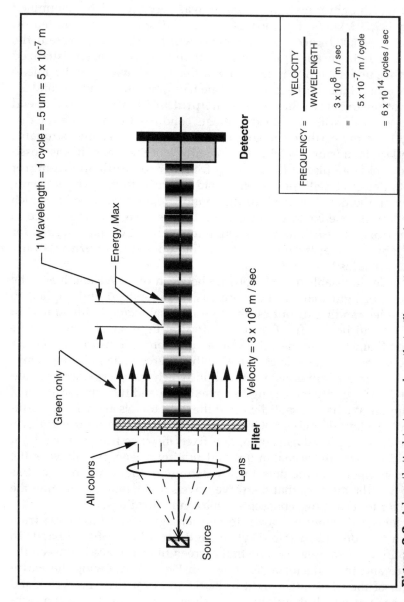

Figure 3-3. A hypothetical test procedure that will permit one to determine the frequency of an emitted wave of energy, knowing the speed at which it travels and its wavelength.

$$\text{Frequency} = \frac{\text{velocity}}{\text{wavelength}}$$

$$= \frac{3 \times 10^8 \, \text{m/s}}{5 \times 10^{-7} \, \text{m/cycle}} = 6 \times 10^{14} \, \text{cycles/s (Hz)}$$

In keeping with our goal, which is to develop a sense for what light is and how it behaves, we now realize that when we see a green light source, the energy from that source is approaching our eyes at a velocity of 3×10^8 m/s, and its energy level is pulsating at a frequency of 6×10^{14} times each second.

As was noted earlier, this discussion has neglected the difference in velocity of light in air as compared with that in a vacuum. The fact is that light in air is slowed by about 0.03 percent, which is indeed negligible for these purposes. A second significant point along these same lines is that measurements have proved the velocity of electromagnetic energy in a vacuum is constant, regardless of its wavelength. In air, the change in velocity with wavelength is so small as to be negligible for our purposes. Therefore, all colors that we see travel at the same speed. It follows, then, that because frequency is inversely proportional to the wavelength, the frequency for each color must be different. In the case of the visual spectrum, that frequency ranges from 7.5×10^{14} for blue light to 6.0×10^{14} for green light to 4.5×10^{14} Hz for red.

Another example, taken from today's technology, will be helpful in establishing our sense for what light is and how it behaves. A common electro-optical instrument of recent years is the laser range finder (LRF). The LRF contains a very pure light source in the form of a laser, a set of projection optics including a shutter, a set of receiving optics, a detector, and a precise timing mechanism. The LRF functions by generating a pulse of laser energy, which is projected to a target where it is reflected back toward the LRF, collected and imaged by the receiving optics onto the detector. The precision timer measures the time required for the pulse to travel to the target and return. With knowledge of the velocity at which light travels, it is then a simple matter to compute the distance to the target. One common LRF configuration utilizes a laser with a wavelength of 1.06 µm and a projected pulse duration of 20 ns. From this information we can generate a realistic description of that pulse including its physical size and characteristics. To generate the pulse the LRF shutter must be opened and then closed in a precise manner, with a total elapsed open time of 20×10^{-9} s. We know that light travels at a velocity of 3×10^8 m/s. Figure 3-4 illustrates the system being discussed. When the shutter has been open for the required 20×10^{-9} s, the leading edge of the pulse will have trav-

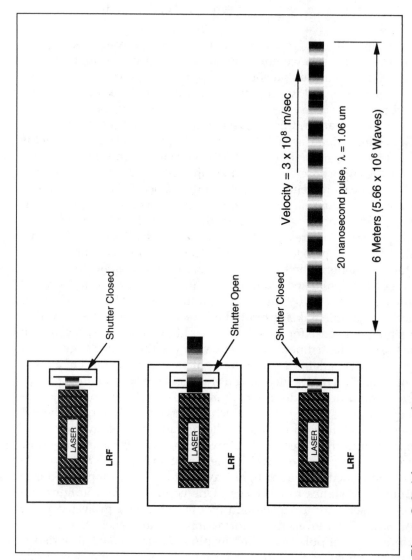

Figure 3-4. A laser range finder generates a short pulse of energy which is used to determine the distance to a target. Shown above is a schematic representation of that pulse.

eled 20×10^{-9} s $\times$ 3×10^8 m/s = 6 m. The shutter will then close, and the result would be a 6-m-long pulse of laser energy, traveling toward the target at a speed of 3×10^8 m/s.

We can determine one other significant characteristic of this pulse of laser energy. We have said that the energy emitted from a laser has a wavelength of 1.06 µm, or 1.06×10^{-6} m. It follows then that the 6-m-long pulse would contain

$$\frac{6.0 \text{ m/pulse}}{1.06 \times 10^{-6} \text{ m/}\lambda} = 5.66 \times 10^6 \ \lambda/\text{pulse}$$

where λ = wavelength.

Returning to Fig. 3-4, we can now visualize and describe the 20-ns pulse from the LRF as a beam of energy, with a wavelength of 1.06 µm, and a total length of 6 m. That pulse contains 5.66×10^6 cycles (wavelengths) of this laser energy, and the entire pulse is traveling through space at a velocity of 3×10^8 m/s.

Exercises such as this are most valuable in the sense that they serve to convert abstract concepts into real-world situations that we can more easily visualize and understand. With that understanding the optical engineer can obviously better generate designs implementing these concepts.

3.4 Wavefronts and Rays

In considering the path that light follows when it leaves a source, things are simpler if we first consider that source to be relatively small, what we will refer to as a *point source.* The light originating at a point source spreads out from that source, forming an expanding spherical wavefront. If we consider a distant source, such as a star, then the radius of that wavefront, as it is detected here on earth, will essentially be infinite and the wavefront will be flat or, as we like to say in optics, *plano.* In most optical engineering cases the light from a source is most conveniently described and dealt with in terms of light rays rather than wavefronts. *Light rays* are lines emanating from the point source and, by definition, traveling perpendicular to the propagated spherical wavefront (see Fig. 3-5).

3.5 Light Sources

A discussion of light sources would seem to involve an endless number of things, but, in reality, is limited to just a few. The sun is obvi-

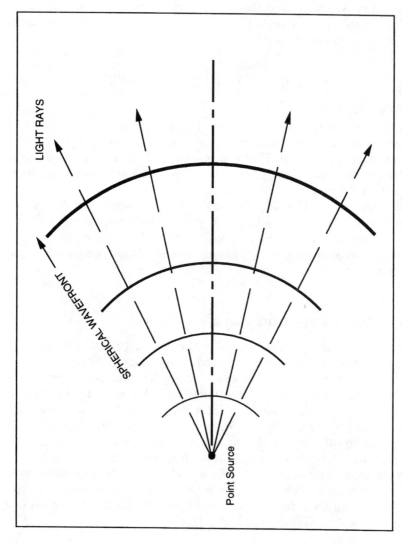

Figure 3-5. As light travels from a point source, it creates a spherical wavefront. It is convenient to assume the existence of light rays as shown, which are radial lines, originating at the source and traveling in a straight line.

ously our most common source of light. Other sources consist primarily of burning fuels and heated filaments. We are able to read the printed words on these pages, not because they are illuminated from within, but because some external source has flooded them with light that has been reflected to our eyes. Likewise with most objects that we view or otherwise optically record during daylight hours, these objects are illuminated by the sun and then reflect a portion of that sunlight to our eyes. In viewing a complex scene we detect varying levels of object brightness as a function of that object's reflectivity at the point being viewed. Likewise, we detect different colors when the object reflects different portions of the spectrum in differing amounts. The grass and leaves reflect primarily the green light and thus appear to be green. Similarly, the red convertible has been painted with a product that reflects the red portion of the visible spectrum while absorbing all other colors.

For most fundamental purposes of optical engineering and analysis, it is valid to treat an object that is reflecting light from a source, as a source itself. In this way we can handle subsequent analysis as if the light originated at that object, then traveled through an optical system and onto a detector for viewing or recording. It is important when doing such analysis to keep in mind that the spectral characteristics of the original source will be modified by the reflectivity characteristics of the object being viewed, resulting in the apparent spectral content of the object. For example, if a white box is viewed in sunlight, it appears white. If viewed at night when illuminated by a sodium arc lamp whose spectral output band is primarily orange, then the white box will appear to be orange.

In a vast majority of cases the object under consideration is being illuminated by one of three common light sources, or illuminants: a tungsten filament lamp, direct sunlight, or average daylight. These sources have been designated "standard illuminants A, B, and C" respectively, by the International Commission of Illumination, for purposes of colorimetry discussions. Figure 3-6 represents a normalized version of the relative spectral output of these three common sources of illumination. These data will often be found useful in determining the spectral nature of an object during the analysis of an optical system.

3.6 Behavior of Light Rays

A light ray is not so much a *thing* as a *concept*. Earlier we defined a light ray as a straight line, originating at a point on the object and extending

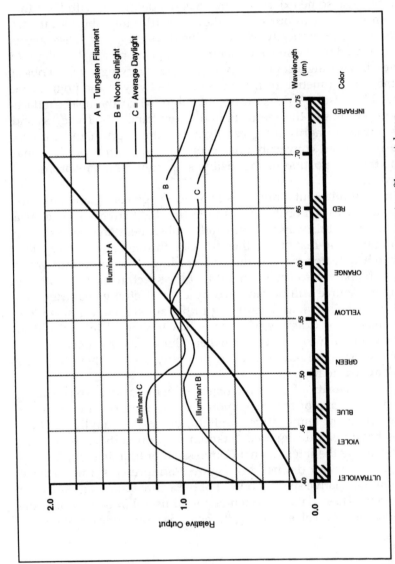

Figure 3-6. Three common light sources, or illuminants: tungsten filament lamp, noon sunlight, and average daylight. These have been designated standard illuminants A, B, and C respectively by the International Commission of Illumination.

to some point on the wavefront that has been generated by that object. The light-ray concept is particularly important in that it makes it much easier for us to visualize and to calculate the behavior of light as it travels from an object, through an optical system and then forms a final image, usually at a detector. A light source does not emit light rays; it emits a spherical wavefront that can be very conveniently represented using light rays. The light ray concept was illustrated in Fig. 3-5.

The value of dealing with light rays as opposed to waves and wavefronts will first be demonstrated by a discussion of reflectors (mirrors) and how they modify the light that is incident on them. The simplest case would be a flat, or plano mirror. In Fig. 3-7, there is a point source of light labeled *object* and, at a distance S to the right, there is a flat mirror. The object may be an original source, such as a lamp, or it may be (and more commonly is) a point on an extended object that is reflecting light from an original source. A second point worth noting is that while the object may be giving off light into any portion of a complete sphere, we are concerned only with that portion of the light from the object that intersects the optics being considered, in this case the mirror.

The objective of this example is to generate a graphic representation of how the light from the object is affected by the mirror, and to allow us to develop an understanding of that behavior with which we are comfortable. The following basic optical principle applies:

The Law of Reflection
When an incident ray of light is reflected from a surface, that incident ray, the normal to the surface, and the reflected ray will fall in the same plane, and the angle of incidence will be equal to the angle of reflection.

In Fig. 3-7, the plane containing the light rays is chosen to be the plane of the paper; thus the angle of incidence i will be equal to the angle of reflection r as shown. Geometric construction (or *doing the math*) will lead to the same conclusion, namely, that an image of the object will be formed on the optical axis (a unique light ray that travels from the object, normal to the mirror surface) at a distance S' to the right of the mirror surface. In the case of a flat mirror, S' will be equal to S.

This very basic exercise illustrates quite nicely the value of the light-ray concept, and how easily it can be applied to the analysis of a simple ray trace problem. By tracing just a single ray from the object it is possible to determine the exact location of the image that will be formed by the mirror.

Optical engineering often involves establishing a basic condition, such as the plano mirror, and then modifying certain factors and

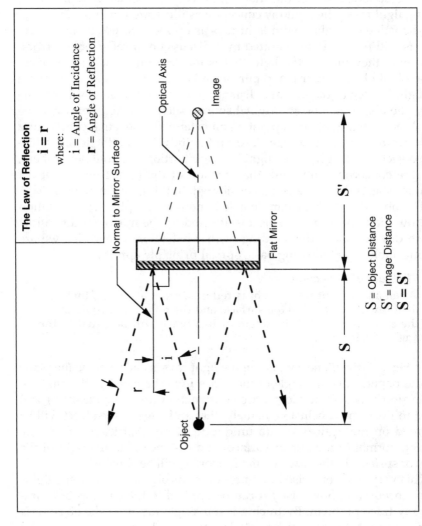

Figure 3-7. Illustrating the use of the light-ray concept to analyze the interaction between light from a point source object, and a plano (flat) mirror.

determining the result. For example, we might be interested in the effect should the distance from the object to the mirror S be modified. Quite obviously in this case each change of S would produce an equal change to S'. Considering a real-world example, when looking into a mirror, if there is some feature on our face that we wish to examine more closely, we move closer to the mirror. The net result is that for each inch we move closer to the mirror, our image behind the mirror moves closer by the same amount ($S = S'$). Thus, the image we are viewing is now 2 inches closer to our eye.

A second variable to the simple mirror example, and one that is much more significant and interesting, is the shape of the mirror surface. Typically, a mirror will be either flat, concave, or convex. When the mirror is assumed to be convex, as in Fig. 3-8, it can be seen that one major resulting change is that the normal to the mirror surface, which in the case of the flat mirror was parallel to the optical axis, is now tilted such that its extension passes through the center of curvature of the mirror. Since the law of reflection remains inviolate, the reflected ray now leaves the mirror surface at a steeper angle than was the case for the flat mirror. The result in this case is that the image is formed closer to the mirror, without the object having been moved at all. Here again, the tracing of a single ray allows determination of image location.

This example explains (in part), the origin of that familiar phrase "Objects in mirror are closer than they appear," which we all have seen etched on the passenger side (convex) mirror on our car. I say "in part" because if it were purely a matter of the image having been moved closer to the mirror, then the objects would actually be *farther* than they appear. What has happened simultaneously is that the convex mirror has reduced the size of the image, by a greater factor than it has moved it closer to the eye. The result is that, while the image is actually closer, it appears considerably smaller. When we see an image of a familiar object (such as a car in the mirror), we judge its distance according to the size of that image. If the object appears smaller, we conclude that it must be farther away...but in this case it is not; it is closer than it appears. Try to think of and accept this example as interesting rather than confusing. We will spend much more time on the subject of object–image relationships and their determination in future chapters. Again, the key objective at this point is to develop a comfortable feel for the behavior of light rays as they pass from an object, reflect from a mirror and create an image.

In Fig. 3-9 we see the same object with a concave mirror at the same distance S. Again, the normal to the mirror surface is represented by a line passing through the center of curvature of the mirror. While this normal was parallel to the optical axis for the flat mirror case, it is now

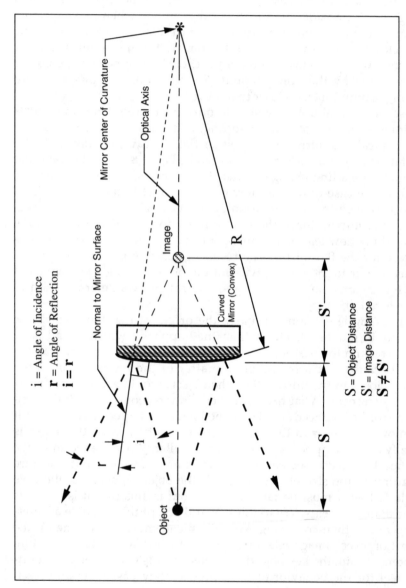

Figure 3-8. Scheme illustrating the use of the light-ray concept to analyze the interaction between light from a point source object, and a curved (convex) mirror.

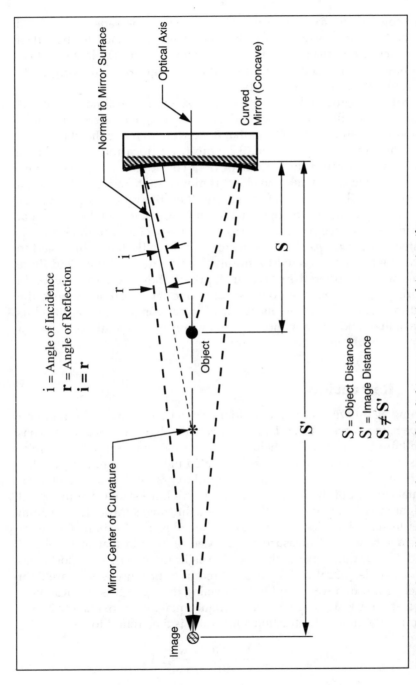

Figure 3-9. Scheme illustrating the use of the light-ray concept to determine the interaction between light from a point source object, and a curved (concave) mirror.

tilted toward the axis. As a result, we see that the reflected ray will also be tilted downward, such that the image is formed to the left of the object at the distance S' from the mirror. Again, the law of reflection has been applied to the tracing of a single ray, allowing us to determine the exact image location.

It will be useful to discuss a special case while the situation involving a concave mirror as shown in Fig. 3-9 is being considered. It is fairly obvious from the illustration that if we were to move the object closer to the mirror by some small amount, the location of the image would move away from the mirror by a considerably larger amount. In the extreme case, when the object is moved to the midpoint between the mirror and its center of curvature, then the reflected ray will be essentially parallel to the optical axis and the image will be formed at an infinite distance from the mirror. In this specific case the source is located at the *focal point* of the mirror and the reflected light is said to be collimated, or projected to infinity. Again, this condition and these terms will be covered in greater depth as they are encountered in future chapters. At this point we are concerned primarily with the light-ray concept and how its application has conveniently allowed us to visualize and understand the interaction between an object and a simple optical component, a mirror.

3.7 Refraction

Starting again with a basic definition, refraction is "the deflection of a propagating wave." The relationship between light waves and rays has been introduced. It will be helpful to now establish a sense for the relationship between a wave and its associated light rays as refraction takes place. In order to do this it will be helpful to first understand the fundamental concept of the index of refraction. We have stated that the speed of light in air is about 3×10^8 m/s. As light travels through a medium other than air, its velocity is reduced. The index of refraction of a material is determined by measuring the speed of light in that material and dividing that number into the speed of light in air. The two most common examples of refraction encountered in optics discussions are light traveling through water and light traveling through glass. Typical measurements for water yield a speed of light equal to approximately 2.25×10^8 m/s. The index of refraction n for water is then found to be

$$n_{water} = \frac{3 \times 10^8 \text{ m/s}}{2.25 \times 10^8 \text{ m/s}} = 1.333$$

In the case of optical glass, there are a number of glass types, all with slightly different index of refraction characteristics. For the case of the most commonly encountered glass type, and a reasonably typical value for most glasses, an average index of refraction of 1.52 represents a valid assumption. From this value we may conclude that the speed of light in a typical optical glass would be 1.97×10^8 m/s. In other words, light leaving a source travels through air at a speed of 3×10^8 m/s, and when this light enters a typical optical glass, the speed at which it is traveling is reduced to 1.97×10^8 m/s. Figure 3-10 shows graphically the relationship between the speed of light in a vacuum, where the index of refraction is 1.00, and the speed of light in a variety of other optical materials where the index is greater than 1.00.

With this relationship between the speed of light and the index of refraction as background, let's now consider the refraction (deflection) of a spherical wavefront when it enters a glass block as is illustrated in Fig. 3-11. This illustration shows a point source of light and the spherical wavefront that is traveling from that source in a left-to-right direction. When this spherical wavefront encounters the glass block shown, it can be seen that the point on the wavefront that is on the optical axis is the first to enter the glass. At this point, that portion of the wavefront is slowed, while that portion remaining outside the glass is not. The net result is that by the time the entire wave is inside the glass, its curvature has been reduced, or flattened. The radius of the wave is now greater than it would have been if the glass were not present. In other words, the wavefront has been deflected, or refracted.

Likewise, the associated rays seen in the figure will bend, or refract when they enter the glass. The amount of this bending may be precisely calculated using the law of refraction, or Snell's law, which states the following:

$$n_1 \sin i = n_2 \sin r$$

where n_1 = index of refraction to the left of the boundary
n_2 = index of refraction to the right of the boundary
$\sin i$ = sine of the angle of incidence
$\sin r$ = sine of the angle of refraction

and the normal at the boundary, the incident ray and the refracted ray, all lie in a common plane.

It is interesting to note that, for the case illustrated, the glass block is shown with its entrance and exit faces flat and parallel. As a result, the amount of refraction for the emerging rays will be equal to the amount

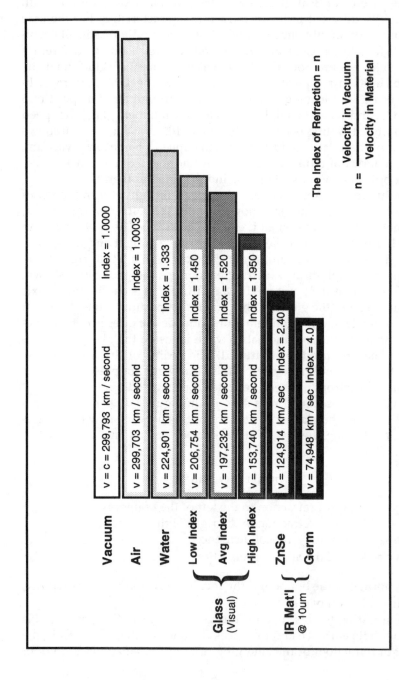

Figure 3-10. The speed at which light travels is a function of the material in which it is traveling. Shown above is the speed and resulting index of refraction for a number of common optical materials.

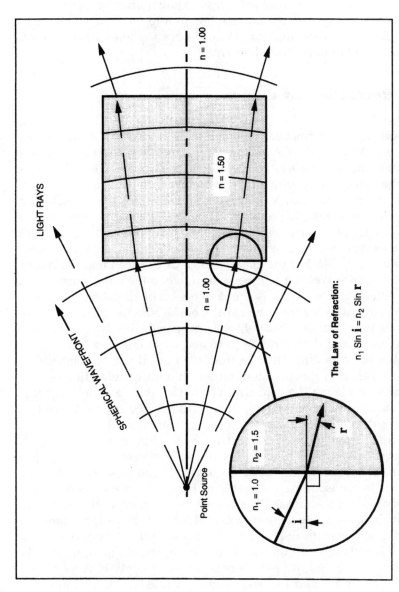

Figure 3-11. As light travels from a point source, it creates a spherical wavefront. When the wavefront enters a block of glass, its shape is changed. The behavior of associated light rays is defined by the *law of refraction*, also known as *Snell's law.*

at entering, and the angle of the ray relative to the optical axis will be restored to its original value. This sense of how refraction occurs when a wave or light ray passes from one medium to the next will be most helpful in the appreciation of numerous optical phenomenon that are to be discussed in upcoming chapters.

3.8 Refraction by a Lens

The case illustrated in Fig. 3-12a is similar to that in 3-11, except the point source of light has been assumed to be located at a great distance to the left. As a result, the incident wavefront is plano (flat) and the light rays from that source are parallel to the optical axis. These rays strike the glass surface with an angle of incidence equal to zero and as a result, are not refracted as they enter and leave the glass block. This case illustrates a lens with zero optical power, or a simple window.

In Fig. 3-12b we have a case where the first surface of the glass block has been modified to have a convex curved surface. As the incident plano wavefront reaches the glass block, it can be seen that the central portion of that wavefront will be slowed, while the outer portion will not be affected. The result is the conversion of the wavefront from plano to curved as it enters the glass. Considering the behavior of the light rays, it can be seen that light rays close to the optical axis have an angle of incidence that is nearly zero and therefore very little refraction takes place. As the distance from the optical axis to the incident ray increases, a proportionately larger amount of refraction occurs, with the result being the conversion of the incident parallel light rays into a bundle of converging rays, indicative of the curved wavefront traveling through the glass. Even though the second surface of the glass block is flat, it can be seen that the emerging wavefront gains additional curvature as it leaves the block. This case (Fig. 3-12b) illustrates the behavior of a classic planoconvex, simple positive lens.

Finally, in the example illustrated in Fig. 3-12c, the glass block has been configured such that its entrance face is now concave. In this case, in a manner quite similar to that of the convex entrance face, the incident plano wavefront is again refracted such that it becomes spherical as it travels through the glass. As the wavefront emerges from the flat side of the block, it again becomes more steeply curved. In this case the wavefront and its associated light rays are diverging rather than converging. This case (Fig. 3-12c) illustrates the behavior of a classic planoconcave, simple negative lens.

This description of refraction and these simple lens forms will be expanded on considerably in subsequent chapters. At the risk of

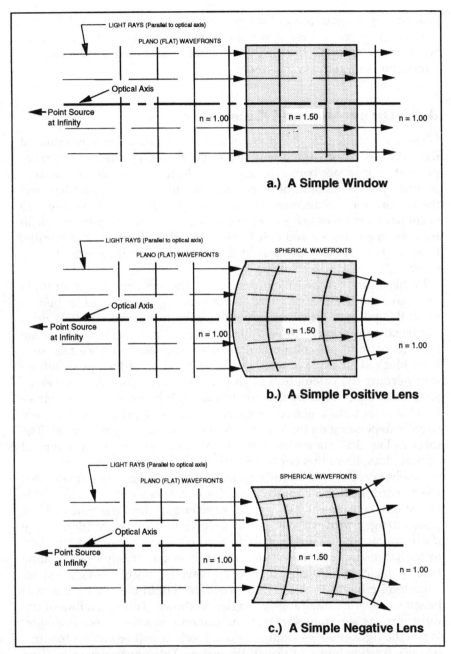

Figure 3-12. As light travels from a distant point source, it creates a plano (flat) wavefront. When that wavefront enters a block of glass, its shape is changed. The amount and nature of that change depends on the characteristics of the glass block.

redundancy, I again remind the reader that the objective at this point has been to present these basic concepts in such a way that the reader will gain a comfortable sense of how light behaves when it encounters a reflecting or a refracting element.

3.9 Dispersion and Color

Dispersion, in the case of light rays, can be defined as the separation of light into its color components by refraction. When refraction takes place at an interface between two materials, it is nearly always accompanied by some amount of dispersion. It will be useful to understand the reasons for this and some of the situations that result. Figure 3-13 is similar to the window example that we have seen earlier, except in this case the entrance and exit faces of the glass block are not parallel to each other. A block of glass in this configuration is referred to as a prism.

In this figure, for the sake of simplicity, the wavefronts and multiple rays have not been shown. Our understanding of light and its behavior at this point tells us that, while these things are present, they need not be shown in every illustration. In Fig. 3-13, a single ray of white light (containing all colors) is shown striking the entrance face of a glass block at an angle of incidence i. Application of the law of refraction permits the calculation of the angle of refraction r. This seems simple enough, but there is a complication. When the index of refraction for most optical glasses is measured precisely, it is found to vary slightly depending on the wavelength of the light being measured. The inset in Fig. 3-13 shows an index vs. wavelength curve for a typical optical glass. From this curve we can see that the index of refraction is somewhat greater in the blue region of the spectrum, while it becomes less toward the red end. Since the index of the incident medium (air) is essentially constant at 1.00 for all wavelengths, and the index of the glass changes with wavelength, it follows that by application of the law of refraction the angle of refraction will also change as a function of the wavelength of the incident light. If in our example we assume that the incident light ray contains all wavelengths of the visual spectrum (white light), then the refracted light within the glass block will be spread over a finite range of angles as shown. This spreading of the light into its separate spectral components is known as *dispersion*. When the light emerges from the glass block, it will be subject to additional refraction and additional dispersion. This phenomenon is often utilized in an optical instrument where separation of the light into its many spectral components is desired. More often, dispersion is found

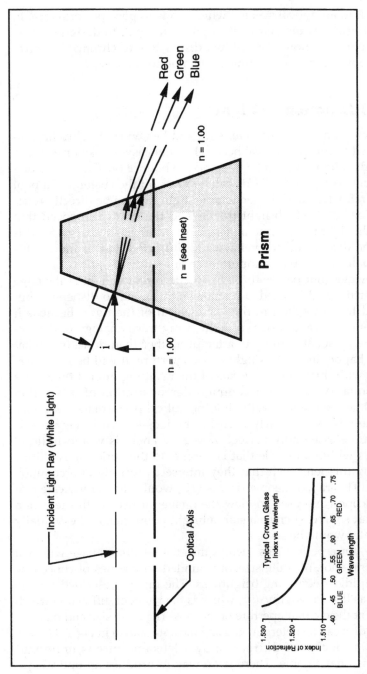

Figure 3-13. A light ray that is incident on a glass block at some angle of incidence i will be refracted as it enters the block. In addition, because the index of refraction of the glass varies with wavelength (see inset), the amount of refraction that occurs will vary as a function of the wavelength of the light. This spreading of white light into its spectral components on refraction is known as *dispersion*.

to be a fault in an optical system, which leads to poor performance. In that case it becomes the task of the optical designer to design the system such that the amount of residual dispersion, or chromatic aberration, is acceptable in terms of final system performance.

3.10 Diffraction of Light

In this discussion of the characteristics and behavior of light, the subjects covered to this point have been relatively easy to visualize and to understand. A light ray can be imagined to bounce off the surface of a mirror, much as a pool (billiard) ball bounces off the cushion of a pool table. The refraction and dispersion of a light ray can be directly related to the speed at which light travels and the change in speed that occurs as the light travels into a glass block from air. In discussing diffraction, the theory and the behavior of diffracted light is more difficult to visualize, but no less important to our study of optics.

When a wavefront passes through an aperture, the light at the edge of that aperture is diffracted, a phenomenon similar to dispersion but not related to a variation in the wavelength of the light. Because it derives from the wave theory of light, a ray trace analogy of diffraction is not a reasonable thing to attempt. Rather than delve into what is actually happening to this light at the aperture, it will be more useful to deal with diffraction in terms of the resulting impact that it has on image quality. The most common demonstration of diffraction effects will be seen when evaluating the optical performance of a lens that is otherwise very nearly perfect. The lens shown in Fig. 3-14 is assumed to be essentially perfect. Ray trace analysis would indicate that all rays within a bundle that is parallel to the optical axis will be refracted by this lens such that they intersect precisely at a common focal point. This ray, or geometric analysis, would further indicate that if a microscope were used to view the image formed by this lens of a distant point source, such as a star, a bright point image of essentially zero diameter would be seen.

In reality, for a perfect lens such as this, the resulting image will be a bright central spot, of finite size, surrounded by a series of concentric rings of rapidly decreasing brightness. The image described here is referred to as the classic *Airy disk*, which is the result of diffraction which takes place at the circular aperture of the lens. The exact size and makeup of the Airy disk can be determined using the data shown in Fig. 3-14.

This basic concept of diffraction by a lens aperture is important because it is unavoidable. There is no way to produce an optical system that will perform better than the limits that are predicted by dif-

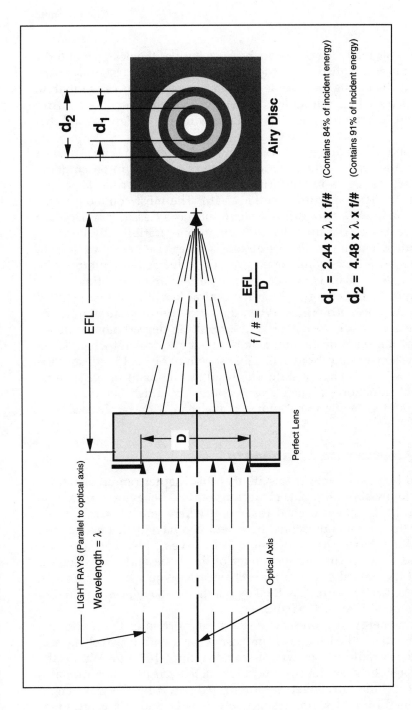

Figure 3-14. A "perfect" lens, one that is completely free of aberrations, would be expected to produce a perfect point image of a point source such as a star. In reality, diffraction effects at the lens aperture lead to the creation of the Airy disk rather than a perfect point image.

fraction. This really is not so bad, because it gives the designer and the manufacturer a realistic target to strive for. Once the design has been corrected (including reasonable manufacturing tolerances) so that all light rays traced fall within the diameter of the central spot of the Airy disk, there is little to be gained by improving the design beyond that point.

A second diffraction effect frequently encountered in optical system design comes about when the lens aperture contains an obscuration. This usually appears in the form of an opaque circular blockage centered on the lens aperture (see Fig. 3-15). The most common occurrence of this is in a two-mirror system, such as a Cassagrain telescope objective. Again, the presence of such an obscuration only becomes important when the basic lens design is otherwise nearly perfect. In such a case the lens would again produce a diffraction pattern in the form of an Airy disk. However, the presence of an obscuration would cause some of the energy within the central bright spot to be shifted out into the concentric rings. When the ratio of the obscuration to the aperture size is between $\frac{1}{3}$ to $\frac{1}{2}$, the resulting degradation of image quality will be significant and must be taken into account when developing system specifications. The table shown in Fig. 3-15 indicates the magnitude of this energy shift. The result, in terms of image quality, will be a reduction in image sharpness and contrast. For these very reasons, systems with obscurations greater than $\frac{1}{2}$ are rarely used.

3.11 Review and Summary

It would be presumptuous to state that this chapter represents all we need to know about light, its characteristics and behavior. Suffice it to say that, if you have reached that point where you are comfortable with all the concepts presented here, and you have developed a good *feeling* for what light is and how it behaves, then you are well equipped to deal with the remainder of this book and many (indeed most) of the optical engineering problems that you may encounter in the future. Let's now review the key points that have been covered in this chapter with regard to light.

Light is energy. It represents just a small portion of the *electromagnetic spectrum*, which ranges from cosmic rays with a wavelength of 10^{-9} µm, to radio waves with a wavelength of 10^{10} µm. Within the electromagnetic spectrum we find the *optical spectrum*, which contains energy from the ultraviolet to the infrared, covering the wavelength band from 0.1 µm to 30 µm, respectively. Finally, and of greatest interest to most of us, within the optical spectrum we find the *visible spec-*

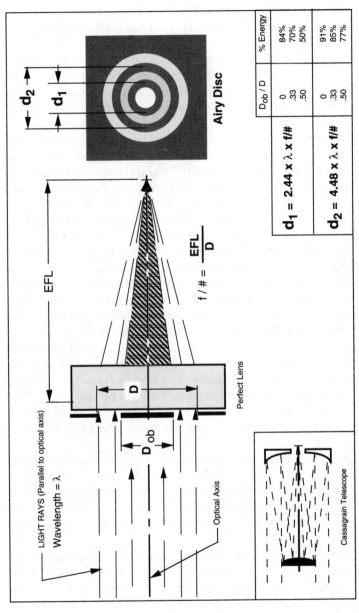

Figure 3-15. Often times an optical system will have a circular aperture with an opaque obscuration at its center. In this case the diffraction effects will result in a reduction of the energy level within the central spot of the airy disk, with that energy shifted outward to the concentric rings. The result will be a significant drop in the contrast of the image.

trum, which contains that energy that we can see, ranging in color from deep violet to deep red, with a corresponding wavelength range from 0.45 to 0.75 μm. In the center of this visible spectrum is the color yellow with a wavelength of 0.56 μm. This is the peak sensitivity point of the human visual system.

Light is energy that travels very fast, so fast that its speed was thought by many, for many years, to be infinite. Over the years, methods have been developed to measure its exact speed. Best measurements to date indicate that speed to be approximately 3×10^8 m/s, or 186,000 mi/s.

When light is emitted from a point source, it may be thought of as traveling in the form of a pulsating, rapidly increasing spherical wavefront. The frequency at which the energy level pulsates is a function of the wavelength of the energy. If the point source is located at a great distance from the optics being considered, it is valid to assume the incident wavefront to be flat (plano). The light-ray concept facilitates our understanding and analysis of the behavior of the wavefront. A light ray is a straight line, originating at the source. The light ray is perpendicular to the wavefront at its point of intersection.

When light encounters an optical component, there are several possible reactions. On striking a mirror, the light will be reflected from the mirror surface. The shape of the mirror (flat, convex, concave) and the law of reflection will dictate the direction taken by a reflected light ray. When incident on a block of glass, the light will enter that glass. The speed at which the light travels inside the glass is dictated by the index of refraction of the glass. The law of refraction permits us to calculate precisely the new direction that will be taken by a light ray after it enters the glass. A curved surface on the glass block will result in a change in the curvature of the wavefront; the resulting component is a lens.

Optical glass is found to have an index of refraction that varies with wavelength. As a result, when a light ray is refracted at an air–glass interface, the degree of refraction will be a function of wavelength. Thus, if the incident ray is made up of white light (all colors), then the phenomenon of dispersion will split that ray into its many component colors.

Finally, we have learned that when a wavefront passes through a limiting aperture, some of the energy is diffracted, or deviated from the path that would have been predicted by conventional ray tracing analysis. This diffraction is present in all systems and when all other limitations have been overcome, the image quality of a perfect optical system will be limited by these diffraction effects.

From among the many facts and figures contained in this chapter, the following are frequently referred to and, as a result, are worth remembering:

Speed of light	3×10^8 m/s
Visible spectrum	0.45 μm (blue), 0.55 μm (green), 0.70 μm (red)
The law of reflection	$i = r$
The law of refraction	$n_1 \sin i = n_2 \sin r$
Index of refraction	$V_{\text{in vacuum}}/V_{\text{in medium}} = 1.33$ for water = 1.52 for average optical glass
Airy disk diameter (first dark ring)	$2.44 \times \lambda \times (f/\#)$

4

Thin-Lens Theory

4.1 Definition of a Thin Lens

A *thin lens* is a lens whose thickness is assumed to be zero and therefore is negligible. The thin lens is a design tool, used to simulate a real lens during the preliminary stages of optical system design and analysis. This concept is particularly valuable because it enables the optical engineer to quickly establish many basic characteristics of a lens or optical system design. By assuming a lens form where the thickness is zero, the formulas used to calculate object and image relationships are greatly simplified. The drawback to the thin-lens approach is that it is not possible to determine image quality without including the actual lens thickness (and other lens data) in the calculations. As a result, while it is possible to establish many valuable facts about an optical system through the application of thin-lens theory and formulas, the ultimate quality of the image can, at best, only be estimated.

4.2 Properties of a Thin Lens

Figure 4-1 is an illustration of a positive thin lens. Any lens or system analysis must start with several known factors which will generally be provided by the end user, or the customer. From these given factors it will be possible, using thin-lens theory and formulas, to generate the missing information required to completely describe the final lens system. In the case shown in Fig. 4-1, for example, it is given that the system detector (image size) will be 25 mm in diameter and that the full

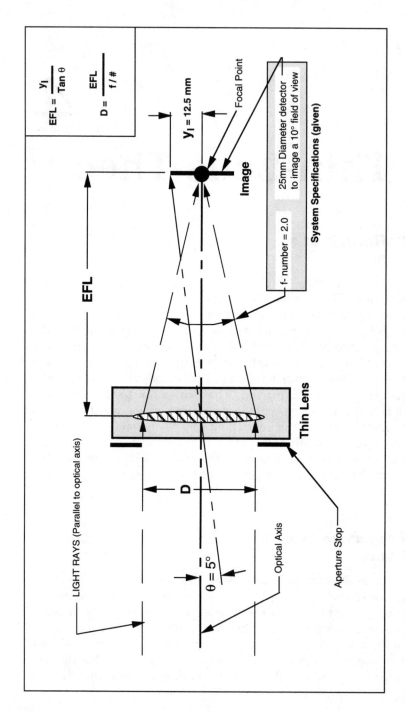

Figure 4-1. Shown are the basic parameters of the thin lens. In the example discussed in the text, the image size y_I, the field of view θ, and the lens speed (f number) are given. From these it is possible to determine the focal length (EFL) and diameter (D) of the lens.

field of view for the system is to be 10°. Other systems considerations dealing with required image brightness indicate that a lens speed (f number) of $f/2.0$ will be required. Applying the formulas shown in Fig. 4-1, since we know the image size and the field of view, we can derive the lens effective focal length (EFL). The image dimension used for this calculation is measured from the optical axis and is designated y_I. In this case the maximum value for y_I is 12.5 mm. Since the half field of view (θ) is 5°, the EFL of the thin lens can be determined using the formula:

$$EFL = \frac{y_I}{\tan \theta} = \frac{12.5}{\tan 5°} = 143 \text{ mm}$$

The diameter of the thin lens is a function of its EFL and f number, based on the following relationship:

$$f \text{ number} = \frac{EFL}{\text{diameter}}$$

$$\text{Diameter} = \frac{EFL}{f \text{ number}} = \frac{143}{2} = 71.5 \text{ mm}$$

Having now established a complete set of thin-lens specifications for this lens system, our next step would be to select a commercial lens that meets those specifications and compare its image quality with the requirements of the system. If not acceptable, then a custom lens design would be in order.

4.3 Aperture Stop, Entrance and Exit Pupils, and Field Stop

Any lens assembly will have one lens aperture, or mechanical component, that limits the diameter of the axial bundle of light (that bundle of light rays originating at the center of the object) that is allowed to pass through the lens to the image plane. For example, in the thin lens shown in Fig. 4-1, a mechanical opening labeled *aperture stop* is shown in front of the lens. It can be seen that without this stop in place, the lens would be able to accept and pass a larger bundle diameter than D. Frequently the aperture stop will take the form of an adjustable iris diaphragm. This allows the effective diameter of the lens to be varied such that the brightness of the image formed at the detector is constant, regardless of the brightness of the scene being imaged.

The entrance and exit pupils of the lens assembly are directly related to the aperture stop. The entrance pupil is the image of the aperture stop as seen when looking from the object side of the lens, while the exit pupil is the image of the aperture stop as seen when looking from the image side of the lens. In the case where the aperture stop is seen directly (without looking through any lenses), then the corresponding pupil will coincide with the aperture stop. The effect and relationship of the entrance and exit pupils to the aperture stop location are illustrated in Fig. 4-2, where the aperture stop location has been changed relative to the lens.

Similar but unrelated to the aperture stop, is the field stop of the lens or optical system. The field stop is the component that limits the field of view seen or imaged by the system. We have stated that for the lens in Fig. 4-1, the detector has a diameter of 25 mm. This detector might be an image intensifier tube or a charge-coupled device (CCD) array within a TV camera. Whatever the case, it is this detector that limits the field of view of the system. Therefore, it is the system field stop.

Were the detector indeed a CCD array within a TV camera, it would follow that the ultimate image produced by the system would not be 25 mm in diameter, but would most likely be a rectangle with a height to width ratio of 3:4 and a diagonal dimension of 25 mm. While not affecting the lens design process significantly, this rectangular format would lead to the specified field of view for this lens being stated as 8° horizontal × 6° vertical, rather than simply 10°. While the system detector is often the field of view limiting component, and as a result the field stop, there are cases where the detector is capable of imaging a larger field of view than the optical system delivers. In such a case the field stop will be a physical surface within the system. That field stop surface must be precisely located at an internal image plane.

4.4   Reference Coordinate System

In discussing and describing optical systems it is important that we become familiar with the basic reference coordinate system that is being used. Figure 4-3 illustrates the most common system, the one that is used throughout this book. It can be seen that there are three mutually perpendicular axes that are designated X, Y, and Z. The Z

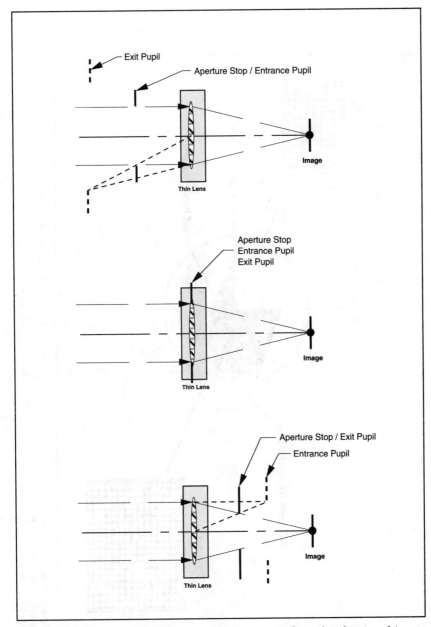

Figure 4-2. The image of the aperture stop as seen from the object and image side of the lens will form the entrance and exit pupils of the lens, respectively. When the aperture stop is in contact with the lens (or nearly so), the entrance and exit pupils will coincide with it (center).

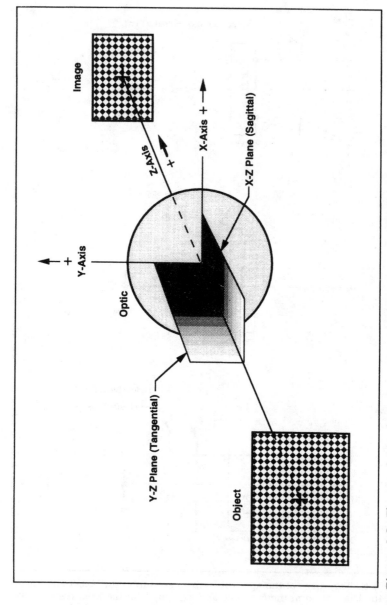

Figure 4-3. The standard reference coordinate system, as used in this and most contemporary optical texts.

axis is also referred to as the *optical axis,* or *centerline* of the system. In most cases, the object, all optical elements, and the image, will be centered on and perpendicular to the Z axis, which is assumed to be horizontal. The Y axis is any vertical line that intersects the Z axis at a right angle. The Y-Z plane is a plane that contains both the Y and Z axes. The Y-Z plane is also referred to as the *tangential plane.* The X axis is any horizontal line that intersects the Z axis at a right angle. The X-Z plane is a plane that contains both the X and Z axes. The X-Z plane is also referred to as the *sagittal plane.*

When it comes to the subject of sign (+ or −), the most important consideration is consistency throughout your work. Generally, and throughout this book, distances along the Z axis that take us farther to the right, or toward the image in the simplest case, are designated positive. In the case of the Y axis, distances away from the Z axis in an upward direction are called *positive.* For the X axis, distances away from the Z axis to the right (as seen from the object) are also called positive. This is the most basic and most common coordinate system. The introduction of mirrors, prisms, tilted, and/or decentered elements will complicate things considerably. This presentation will be limited to this most simple and familiar configuration.

4.5 Thin-Lens Power

The power of a lens is an indicator of how much that lens will converge or diverge an incident wavefront. Lens power is usually indicated by the symbol ϕ, and may be specified in several ways. Most commonly, lens power is given as the reciprocal of the EFL. For example, the power of the lens in Fig. 4-1 would be $\phi = \frac{1}{143} = 0.007$. A second method of lens power specification, used most often for camera close-up lenses and for eyeglasses, is to specify the lens power in diopters. This is simply equal to the reciprocal of the EFL, with the EFL being designated in meters. For the example, when the EFL is 143 mm, or 0.143 m, the lens power might also be specified as 1/0.143 m = 7.0 diopters.

When several thin lenses are arranged such that their separations are negligible, then the powers of those lenses may be added directly. This is a very useful characteristic, considering the fact that it is not possible to determine the focal length of a combination of thin lenses by directly adding the individual lens focal lengths. This is illustrated in Fig. 4-4, where two cases are shown. When there are two positive

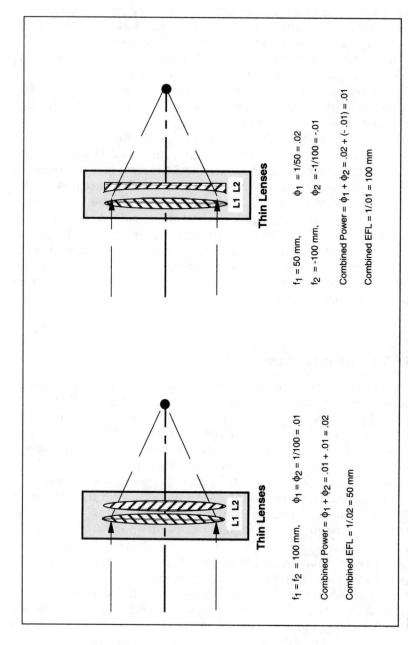

Figure 4-4. Thin-lens power ϕ is a valuable concept in that it permits the combined power and focal length of a combination of thin lenses to be easily determined.

thin lenses in close proximity, each with a focal length of 100 mm, the focal length of the combination can be found by adding their powers and then taking the reciprocal of that sum. In this case the power of each lens would be $\phi = \frac{1}{100} = 0.01$. The sum of the two would be $\phi_{combined} = 0.01 + 0.01 = 0.02$, and the combined focal length would be $1/0.02 = 50$ mm.

The second case (right panel) shown in Fig. 4-4 deals with a common situation where the two thin lenses have powers of different sign. In this case we have a thin positive lens with a focal length of 50 mm and a thin negative lens with a focal length of -100 mm. Adding the two lens powers, we have $\phi_1 + \phi_2 = 0.02 + (-0.01) = 0.01$ for the combined power of the two lenses. The resulting focal length in this case would be

$$\frac{1}{\phi_{combined}} = \frac{1}{0.01} = 100 \text{ mm}$$

Things get a little more complex when the space between the thin lenses is increased. Dealing with the lens powers, the following formula can be used:

$$\phi_{combined} = \phi_1 + \phi_2 - (d\,\phi_1\,\phi_2)$$

where ϕ_1 and ϕ_2 are individual lens powers and d is lens separation.

In a case of this type, it is often simpler to use the formula that deals directly with the EFLs of the two lenses and the separation d between them. For example, recalling the lens that was described in Fig. 4-1, the system specifications that were given called for a 10° field of view and a lens speed of $f/2.0$. An experienced optical engineer might conclude from these numbers that a Petzval lens form would be required in order to assure adequate image quality. A Petzval lens can be represented by two positive thin lenses separated by a substantial airspace. In the classic Petzval layout, the EFL of the first element is twice the EFL of the combination, while the EFL of the second element and the lens separation are both equal to the EFL of the combination. Since we have established the need for a lens whose combined EFL is equal to 143 mm, it follows from the preceding description that the EFL of the first thin lens should be 286 mm, while the separation between the lenses and EFL of the second thin lens should both be equal to 143 mm. This is the arrangement that is illustrated in Fig. 4-5. In addition to the power-based formula given above, Fig. 4-5 contains two formulas that are based on using the individual lens EFLs. The first of these permits the direct calculation of the effective focal length (EFL) for the

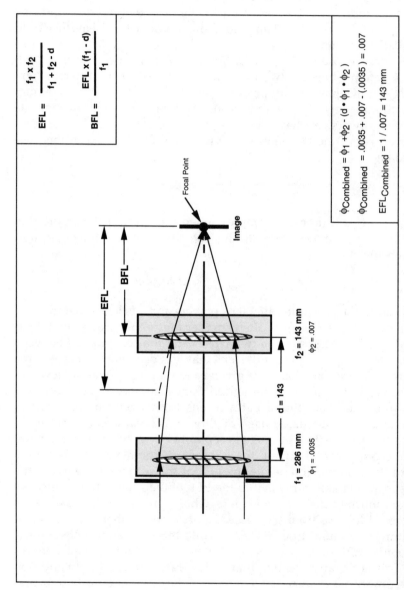

$$EFL = \frac{f_1 \times f_2}{f_1 + f_2 - d}$$

$$BFL = \frac{EFL \times (f_1 - d)}{f_1}$$

Focal Point

Image

BFL

EFL

$f_2 = 143 \text{ mm}$
$\phi_2 = .007$

$d = 143$

$f_1 = 286 \text{ mm}$
$\phi_1 = .0035$

ϕCombined $= \phi_1 + \phi_2 - (d \cdot \phi_1 \cdot \phi_2)$

ϕCombined $= .0035 + .007 - (.0035) = .007$

EFLCombined $= 1 / .007 = 143 \text{ mm}$

Figure 4-5. When two thin lenses are separated by a substantial distance *d*, the focal length of the combination (EFL) and the back focal length (BFL) can be calculated using the formulas shown here.

lens combination; the second permits determination of the distance from the second lens to the image, or the back focal length (BFL). Given the focal length of the two thin lenses f_1 and f_2 and the distance d between them, the following calculations yield the combined EFL and the BFL:

$$\text{EFL} = \frac{f_1 \times f_2}{f_1 + f_2 - d}$$

$$= \frac{286 \times 143}{286 + 143 - 143} = 143 \text{ mm}$$

$$\text{BFL} = \frac{\text{EFL} \times (f_1 - d)}{f_1}$$

$$= \frac{143 \times (286 - 143)}{286} = 71.5 \text{ mm}$$

4.6 Ray Tracing a Thin Lens (Object at Infinity)

The simplicity of the thin-lens concept makes ray tracing through a thin lens equally simple. This ray tracing may be done either graphically or mathematically, with equal ease and effectiveness. First, the optical axis represents the path of a ray traveling from the center of the object to the center of the image. There are just two additional rays that need to be traced in order to establish all the thin-lens characteristics that will be required. These two rays are commonly designated the *marginal* and the *chief* rays. The marginal ray is the ray that originates at the center of the object (where the optical axis intersects the object) and passes through the outermost edge of the optical system's aperture stop. The chief ray is the ray that originates at the edge of the object and passes through the center of the optical system's aperture stop. In the case shown in Fig. 4-6, the object is assumed to be at infinity. As a result, all rays originating at the center of the object (including the marginal ray) may be drawn parallel to the optical axis. The chief ray is the ray that passes through the center of the aperture stop at the maximum specified field angle of 5°. In setting up a single thin lens for graphic ray tracing it should be assumed that the aperture stop is in coincidence with the lens, unless it has been otherwise specified. The focal length of this thin lens has been determined earlier to be 143 mm (see Fig. 4-1). The ray trace layout should include the focal point of the

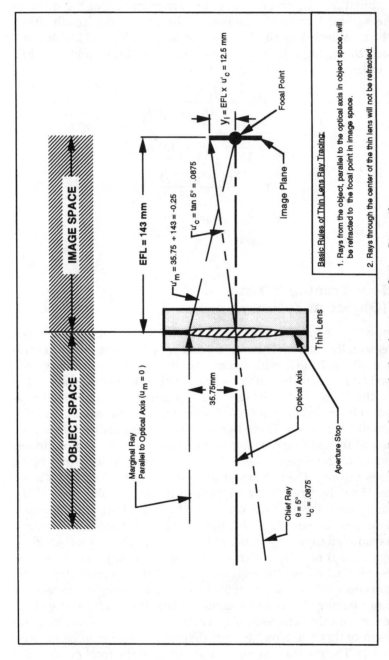

Figure 4-6. Ray tracing through a thin lens with the object at infinity. Given the EFL, half field angle and aperture size, the marginal and chief rays can be traced as shown to determine other thin-lens characteristics. Ray tracing may be done graphically or by the application of simple trigonometric formulas.

thin lens. For this example that would be a point on the optical axis, 143 mm to the right of the lens. By definition, all light rays from the center of an object at infinity (including the marginal ray) will be refracted by the thin lens such that they pass through the focal point. The height of the marginal ray at the lens will be equal to one half the aperture stop diameter. Again, we have previously established the aperture stop diameter to be 71.5 mm. The height of the marginal ray at this point is equal to the aperture stop radius, which is $71.5/2 = 35.75$ mm. The angle that the marginal ray makes with the optical axis as it travels from object to image is important. Ray angles in object space are traditionally designated as u, while ray angles in image space, following refraction by the thin lens, are designated as u'. From Fig. 4-6 it is obvious that the value of u_m for the marginal ray in object space is 0, while for the same ray in image space, its slope angle is

$$u'_m = 35.75 \div 143 = 0.25$$

While it might seem reasonable to convert this value (0.25) to its equivalent in degrees, that will not be necessary, since all subsequent ray tracing and calculations will use the value in its present form. When, as is most often the case, the image is formed in air, the slope angle (u') of the marginal ray in image space is also referred to as the *numerical aperture* (NA) of the thin lens. This is a term used most often when describing the characteristics of finite conjugate optics, such as a microscope objective. A familiar formula, worth remembering, relates the NA to the f number of the lens:

$$f \text{ number} = \frac{1}{2\,\text{NA}}$$

In this case,

$$f \text{ number} = \frac{1}{2 \times 0.25} = 2.0$$

Continuing with the thin-lens ray trace, the *image plane* is a plane that is perpendicular to the optical axis, located at the focal point. The size of the image is determined by the intersection of the chief ray with the image plane. Because we have assumed the aperture stop to be in coincidence with the thin lens, the chief ray will pass through the center of the thin lens at the optical axis. Any ray passing through the center of a thin lens will not be refracted; that is, its angle with the optical axis will not be changed. In the case of Fig. 4-6, the chief ray is travel-ing from the object at an angle of 5° with the optical axis, thus the angle u_c for the chief ray equals the tangent of 5° = 0.0875. This angle is constant for both the object and image side of the thin lens. The height of the chief ray at the image plane may then be found by multi-plying its slope angle u'_c by the EFL of the thin lens:

$$y_1 = 0.0875 \times 143 \text{ mm} = 12.5 \text{ mm}$$

The following two basic rules for thin-lens ray tracing are very use-ful, easy to work with, and easy to remember:

1. A ray that is parallel to the optical axis in object space will be refracted by the thin lens to pass through its focal point.
2. A ray that passes through the center of a thin lens will not be refracted. Its angle with the optical axis will be the same in both object and image space.

It is important to establish a sign convention for ray tracing and to maintain that convention consistently throughout the ray trace analy-sis. The reference coordinate system and sign convention shown in Fig. 4-3 are applicable. All thin-lens ray tracing is done with rays that lie in the Y-Z (tangential) plane. The sign of the angles u and u' will be positive when the value of Y is increasing and negative when it is decreasing. Consider the angle of the chief ray as it appears in Fig. 4-6. At the far left, in object space, the Y value of the chief ray is negative (about −10 mm). At the lens, this Y value has become zero. Since zero is greater than −10 mm, the Y value has been increasing; thus the sign of the 0.0875 chief ray slope angle is positive. For the marginal ray, the Y value at the lens is 35.75 mm. When the marginal ray reaches the image plane, its value is zero. Since zero is less than 35.75, the Y value has been decreasing; thus the sign of the 0.25 marginal ray slope angle is negative. A simple sketch of the lens system will be extremely help-ful in visualizing these ray paths.

4.7 Ray Tracing a Thin Lens (Finite Object Distance)

When the object distance becomes finite, as shown in Fig. 4-7, the thin-lens ray trace becomes a little more complicated. Because the

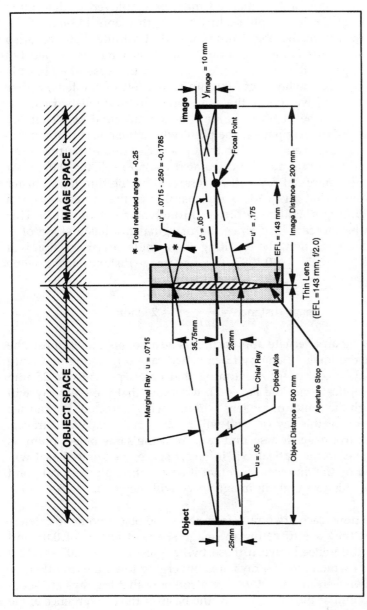

Figure 4-7. Ray tracing through a thin lens with a finite object distance. The marginal ray may be traced from object to image as shown to determine the image distance. In an alternate approach, the chief ray and a second ray from the same object point parallel to the optical axis in object space may be traced as shown, yielding the same result.

71

marginal ray (that ray originating at the center of the object and passing through the edge of the aperture stop) is not parallel to the optical axis in object space, a new method must be employed to determine the location of the image plane. In Fig. 4-7, the same 143-mm, $f/2.0$ lens is shown, with the object now located at a finite distance. Also shown in this figure is the focal point of the lens, which is located 143 mm to the right of the lens in image space. In this case, the object is located at a finite distance of 500 mm to the left of the lens, rather than at infinity. If we follow the same procedure established earlier, the first step will be to trace the path of the marginal ray. That ray will, by definition, originate at the point where the object plane intersects the optical axis and travel to the upper edge of the aperture stop. The angle of the ray u may be calculated as $35.75 \div 500 = 0.0715$. At this point, the marginal ray will be refracted by the thin lens. Earlier we determined that an incident ray, parallel to the optical axis at a height of 35.75 mm, will be refracted through an angle of -0.25. This will again be the case, but since the marginal ray has a slope of $+0.0715$ before refraction, the resulting slope angle after refraction will be $u' = 0.0715 - 0.25 = -0.1785$. The new image distance may now be calculated as

$$\text{Image distance} = \frac{35.75}{0.1785} = 200 \text{ mm}$$

Considering an alternate approach, it is also possible to establish the location of the image plane by tracing two rays originating at a common off-axis object point. Assuming that object point to be 25 mm below the optical axis (see Fig. 4-7), we know that the chief ray will pass through the center of the lens without being refracted. We can see that the angle of this ray ($u = u'$) will be $25 \div 500 = 0.05$ as it travels from object to image. If a second ray, from the same object point, is traced parallel to the optical axis in object space, we know that it will be refracted by the thin lens such that it passes through the focal point of the lens. This ray angle in image space will then be $u' = 25 \div 143 = 0.175$.

We have now traced two rays from the same object point and determined that they are traveling in image space at angles of 0.05 and 0.175. The difference between these two angles ($0.175 - 0.05 = 0.125$) is the rate at which the two rays are converging toward each other, or the angle between the two rays. We also know that the rays are separated by 25 mm at the thin lens. On the basis of the convergence angle we can calculate that these two rays will intersect at a distance equal to $25 \div 0.125 = 200$ mm, to the right of the lens. This intersection point establishes the location of the thin-lens image plane and coincides with the value that was found earlier by tracing the marginal ray.

Having determined the object and image locations, we may now determine the actual size of the object and image. We had arbitrarily chosen an object height of -25 mm, and, through ray tracing the chief ray, we can see that this will produce an image height of $200 \times 0.05 = 10$ mm. If the object is a 50-mm-diameter illuminated disk, then the image will be a corresponding disk with a diameter of 20 mm. From these data we can conclude that the magnification of this thin lens is $20 \div 50 = 0.4 \times$.

If, on the other hand, we assume the earlier condition where the image plane contains a detector with a diameter of 25 mm, then we can conclude that a disk at the object plane that is 25 mm $\div 0.4 = 62.6$ mm in diameter, will be imaged by the lens onto that detector. This 62.6-mm-diameter object disk is referred to as the *field of view* of the thin-lens system.

The magnification factor can also be found by comparing the ratio of the image distance to the object distance ($200 \div 500 = 0.4 \times$). A less direct, but often useful method of determining magnification involves comparison of the marginal ray slope angle in object and image space ($0.0715 \div - 0.179 = - 0.4 \times$). In this case, the minus sign correctly indicates an inversion of the image relative to the object.

To permit calculation of image brightness, it is necessary to know the f number of the light bundle being focused at the image plane of the system. In this case, that f number can be found by dividing the image distance by the exit pupil diameter:

$$f \text{ number} = \text{image distance} \div D_{\text{exit pupil}}$$

$$= 200 \div 71.5 = f2.8$$

We could also employ the formula established earlier, using the marginal ray angle in image space, or the numerical aperture:

$$f \text{ number} = \frac{1}{2\,\text{NA}}$$

$$= \frac{1}{2 \times 0.179} = f2.8$$

Either approach works equally well and should produce the same result. Whenever two approaches are available, it is wise to employ both methods, to check for any possibility of error. It is interesting to note that for this case, where we have a finite object distance, even though we are still dealing with an $f/2.0$ lens the f number at the image plane has been changed to $f/2.8$.

4.8 Rounding Off

It may have been noticed throughout this chapter that numerical results are frequently rounded off, in what may appear at first to be a random fashion. Legitimate rounding off has been done in order to clarify and simplify the presentation, but it has not been done randomly. It is important to always maintain a degree of precision that is consistent with the requirements of the optical system being analyzed. In nearly all cases, the six-place calculations can and should be left to the computer. A rule of thumb, that has served me well in most cases, has been to round off such that the error introduced is significantly less than 1 percent. For example, at the beginning of this chapter we were given a detector size of 25 mm and a required field of view of 10°. Using the maximum image height of 12.5 mm and the half-field angle of 5°, we find the EFL of the thin lens to be 142.876 mm. This was rounded up to 143 mm to simplify subsequent calculations. The error introduced by doing so was less than 0.1 percent. A reality check at this point would tell us that if we were to purchase or manufacture a typical 143-mm lens, the standard tolerance on its focal length would be in the 1 percent range.

In the calculations shown in Fig. 4-5 the individual lens powers are found to be 0.0035 and 0.007. Here, one might be tempted to round the first number off to 0.004, until a quick calculation reveals that this introduces a 14 percent error...not acceptable. In Fig. 4-6 the initial value for the 5° half-field angle is given as 0.0875. This might have been rounded to 0.088, but it is such a basic number to the system (tan 5°) that it was not judged to be a wise move. When it comes to rounding off, hard and fast rules are not possible, nor are they desirable. The decision should always be based on a sound reason. If a sound reason is not apparent then, do not round off. The type of conflicting results that occur due to rounding off can be seen in Figs. 4-7 and 4-8. The first calculation of image distance resulted in a value of 200 mm. Knowing the object distance ($S = 500$ mm) and the thin-lens EFL ($f = 143$), we can use the formula in Fig. 4-8 to calculate the exact image distance:

$$S' = \frac{f \times S}{S - f} = 71{,}500 \div 357 = 200.28 \text{ mm}$$

The error between these two results is 0.14 percent, a negligible and acceptable rounding-off error.

We should also keep in mind the fact that these are all thin-lens calculations, which have been introduced as approximations in their own

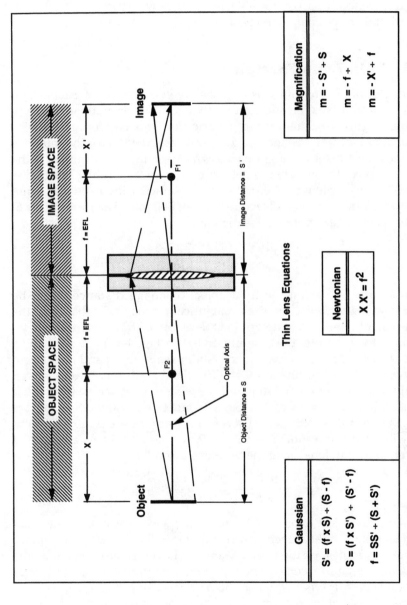

Figure 4-8. In lieu of graphical or mathematical ray tracing, the formulas given here can be used to determine the thin-lens parameters, including object and image distances and magnification factors.

right. To attempt to accomplish an unusual degree of precision here would represent a waste of time and energy. Better that time and energy be devoted to assuring that the proper formulas are being used, and in developing a comfortable sense of how the lens will perform and what its basic characteristics will be.

4.9 Thin-Lens Formulas

Another approach to thin-lens analysis involves the use of established, traditional thin-lens formulas rather than graphic or mathematical ray tracing. Figure 4-8 illustrates a generic thin-lens configuration, with the object at a finite distance. In a typical case, the EFL of the thin lens and the object distance might be known. Applying the formulas in the left-hand box, it is possible to determine the corresponding image distance. The example we used earlier for ray tracing involved a 143-mm EFL (f) thin lens, with an object distance of 500 mm. The formula for S' can be used to determine image distance:

$$S' = \frac{f \times S}{S - f} = \frac{143 \times 500}{500 - 143} = \frac{71,500}{357} = 200 \text{ mm}$$

On another occasion the total object to image distance might be known, along with the required magnification factor. In that case, the requirement would be to determine the required thin-lens focal length. Assume, for example, the case illustrated in Fig. 4-9. Here we have a system with a 35-mm transparency (slide) as an object and a rear projection screen at the image plane. The distance from slide to screen (total track) is fixed at 500 mm and we wish to produce an image that is magnified by a factor of 9 ×. Since S' must be equal to $9S$, it follows that $S+S' = 10S = 500$ mm. Therefore, S must be equal to 50 mm. Knowing that $S = 50$ and $S' = 9S = 450$, we can find the focal length of the required thin lens using the thin-lens formula:

$$f = \frac{SS'}{S + S'} = \frac{22,500}{500} = 45 \text{ mm}$$

Knowing the object to be 24 × 36-mm slide, we can also determine that the image size on the screen will be nine times that, or 216 × 324 mm.

Those formulas, dealing with S and S', have become known as the *gaussian thin-lens formulas*. The equation shown in the center box in Fig. 4-8 ($XX' = f^2$), is known as the *newtonian* thin-lens formula. This formula is easier to remember and is also somewhat simpler algebraically. It should be noted that in the newtonian formula the terms X and X' are distances along the Z axis, from the lens focal point to the

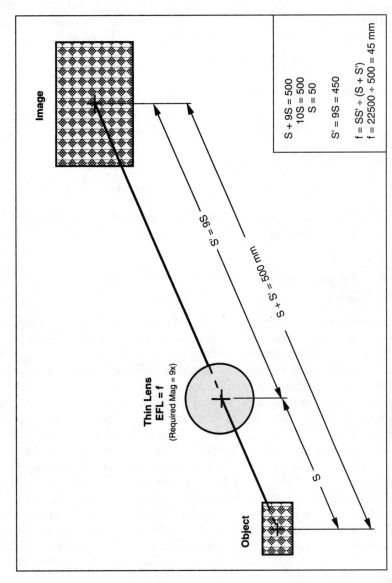

Image

Thin Lens
EFL = f
(Required Mag = 9x)

Object

S

S' = 9S

S + S' = 500 mm

$S + 9S = 500$
$10S = 500$
$S = 50$

$S' = 9S = 450$

$f = SS' \div (S + S')$
$f = 22500 \div 500 = 45$ mm

Figure 4-9. Diagram showing the application of thin-lens formulas to determine the required thin-lens EFL.

object and image. In this case, the use of the term X is based on tradition, and it is in no way related to the X axis, as it has been defined in our local coordinate system.

The newtonian formula can be used to solve the example discussed earlier, where the lens has a focal length f of 143 mm and the object distance S is 500 mm. Using these numbers, we can compute the value for X:

$$X = S - f = 500 - 143 = 357 \text{ mm}$$

Knowing f and X, we can now compute X':

$$X' = f^2 \div X = \frac{143^2}{357} = 57 \text{ mm}$$

The new value for S' will then be

$$S' = f + X' = 143 + 57 = 200 \text{ mm}$$

This confirms earlier calculations, done using the gaussian formulas and thin-lens ray tracing.

4.10 Applications of Thin-Lens Theory

A few examples will be helpful in demonstrating the value and usefulness of thin-lens theory. Assuming once again our basic 143-mm focal length, $f/2$ lens, the question might arise as to over what range of object distances this lens would provide satisfactory image quality. This question generates an immediate need for a more precise definition of the term *satisfactory*. We might be told that any conditions where the blur spot size due to error in focus is less than 0.02 mm in diameter would be considered satisfactory. For the purpose of this example we will ignore the lens aberrations and assume that the lens forms a perfect image at the thin-lens point of focus. For an object at infinity, this perfect image will be formed at the focal point of the lens (see Fig. 4-10). As the object moves from infinity, closer to the lens, the point of best focus moves to the right of the focal point. If our detector is fixed at the focal point, it follows that this change in focus will result in a blur spot at the detector. Because this is an $f/2$ light bundle, that blur spot will reach a diameter of 0.02 mm when the amount of defocus reaches 0.04 mm.

The object distance that corresponds to this 0.04-mm focus shift can be found using the newtonian thin-lens formula:

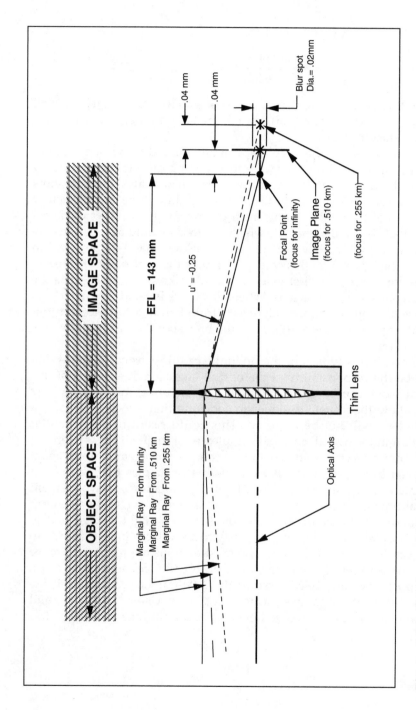

Figure 4-10. Thin-lens ray trace to determine the range of object distances over which satisfactory image quality will be produced. The midfocus object distance (in this case 0.51 m) is referred to as the *hyperfocal distance.*

$$XX' = f^2$$

$$X = \frac{f^2}{X'} = \frac{143^2}{0.04} = 511{,}225 \text{ mm} = 0.51 \text{ km}$$

From this we conclude that, if the lens is focused at infinity, then all objects from a distance of 0.51 km (about ⅓ of a mile) to infinity will be in satisfactory focus.

There is one additional step that can and should be implemented in a case of this type. If the lens were focused at a distance of 0.51 km by moving the detector 0.04 mm to the right, then analysis would show that all objects from 0.51 km to infinity would still be in satisfactory focus. In addition, there would be a certain range of object distances inside the 0.51-km distance for which the focus would also be satisfactory. Repeating the above calculations, we find that a value equal to one half the 0.51 km, in this case 0.255 km, is the near point of satisfactory performance. The first value found (0.51 km) is referred to as the *hyperfocal distance* for this lens. When lens focus is set at the hyperfocal distance, the lens will image all objects from half that distance to infinity with satisfactory resolution, i.e., the blur spot diameter will be less than 0.02 mm.

This exercise is typical of the preliminary thin-lens calculations that might be the responsibility of the optical engineer. The results of this type of work can be quite valuable. For instance, in this case they might allow the system designer to conclude that a focus mechanism for the lens will not be required. This could result in a substantial reduction in the overall cost and complexity of the final lens system.

Consider a second, somewhat more complex problem that can again be solved by applying the basic thin-lens principles that have been presented earlier in this chapter. Figure 4-11a shows a two-element Petzval-type thin lens, similar to the one shown in Fig. 4-5. This example will deal with the question of how that lens might be designed so that it is capable of being continuously focused from infinity, down to an object distance of 2000 mm. The most obvious approach is to make the space between the lens and the detector (image) adjustable. For a first approximation, we can assume the 2000-mm object distance is the X quantity in the thin-lens formula: $XX' = f^2$. Using that value, and knowing the f of the lens to be 143 mm, we can find the value for X':

$$XX' = f^2$$

$$X' = \frac{f^2}{X} = \frac{143^2}{2000} = 10.2 \text{ mm}$$

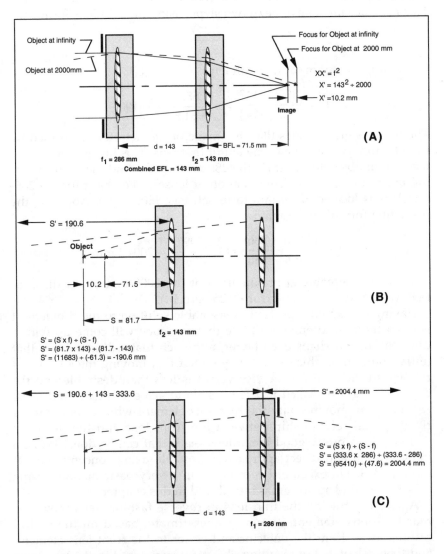

Figure 4-11. Application of thin-lens formulas to determine focus for a near object in a Petzval-type lens. The approximate value of 10.2 mm is found in *a* using the thin-lens formula. Panels *b* and *c* show the thin-lens ray trace, which confirms the adequacy of that approximation.

It is possible to reverse the lens layout, introduce this 10.2-mm increase in the BFL, treat the detector (image) as the object, and then trace rays from that surface to the new image plane (see Fig. 4-11b). This ray trace must be done in two steps, using the thin-lens formula:

$$S' = \frac{S \times f}{S - f}$$

First, for the 143-mm focal length lens:

$$S' = \frac{81.7 \times 143}{81.7 - 143} = \frac{11{,}683.1}{-61.3} = -190.6$$

The minus sign indicates that the image of the detector, as formed by the 143-mm lens, will be to the left of that lens. This image can now be used as an object to ray-trace the 286-mm lens to find the corresponding image plane for the combination of lenses. The object for the 286-mm lens is located 333.6 mm to its left (see Fig. 4-11c). Applying the preceding formula once again, we have

$$S' = \frac{(333.6 \times 286)}{333.6 - 286} = \frac{95{,}409.6}{47.6} = 2004.4 \text{ mm}$$

This is an acceptable approximation of the 2000-mm object distance that was our design goal (error <0.25 percent).

Having generated a perfectly reasonable solution to the problem, it is often the optical engineer's fate that someone will come up with a different and, perhaps even better, approach that will invalidate that initial solution. In this case, it may be felt that moving the entire lens relative to the detector (or vice versa) is less than desirable and the engineer might be asked if it is possible to achieve the desired focus capability by moving only the rear lens element while maintaining a fixed spatial relationship between the front lens and the detector. Sounds like a pretty good idea...let's see if that can be done. While a more elegant and direct approach is possible using concepts not yet presented, I will execute this example in a very basic manner, using only those thin-lens formulas introduced in this chapter.

Again, ray tracing the thin lens in reverse fashion, let's move the rear thin-lens element 10 mm (a guesstimate, based on the earlier results) away from the detector and closer to the front lens element, and then calculate the resulting object distance (see Fig. 4-12a). Using the now familiar thin-lens formula, we can compute the location of the image of the detector as formed by the rear thin lens:

$$S' = \frac{S \times f}{S - f} = \frac{81.5 \times 143}{81.5 - 143} = \frac{11{,}654.5}{-61.5} = -189.5$$

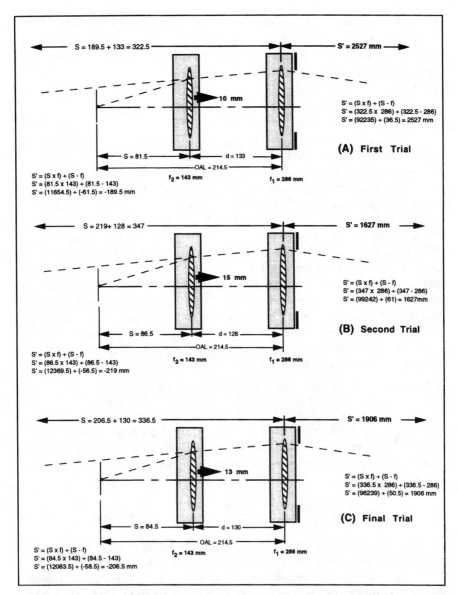

Figure 4-12. It is possible to adjust the focus of the lens assembly by moving only the rear element, while maintaining a fixed relationship between the front element and the image plane. Shown here are the three steps discussed in the text to determine the total amount of lens travel required to focus from infinity, down to a distance of 2000 mm.

This image distance to the detector can now be added to the lens separation and the sum can be used as the object distance S for the front thin lens. We can now compute the distance to its image using the same formula:

$$S' = \frac{S \times f}{S - f} = \frac{322.5 \times 286}{322.5 - 286} = \frac{92{,}235}{36.5} = 2527 \text{ mm}$$

This tells us that, while our first guess as to how far the rear lens must be moved was close, it was not moved quite enough. The lens shift can be changed to 15 mm, and the preceding calculations quickly repeated, yielding a final image distance of 1630 mm (see Fig. 4-12b).

We have now surrounded the required answer (2000 mm) and a linear estimate as to the correct lens shift will probably work. In going from 10- to 15-mm lens travel, our focus distance went from 2527 to 1630 mm. If we want to stop at 2000 mm, that would be about 59 percent of the way through the total lens travel, which would put us at about 13 mm. Let's repeat the calculation once more, assuming the rear lens has been moved away from the detector and toward the front lens by 13 mm. This would reduce the thin-lens separation from an original value of 143 down to 130 mm. The new object distance for the rear lens will be 71.5 + 13 = 84.5 mm. The thin-lens calculations would then be as follows:

$$S' = \frac{S \times f}{S - f} = \frac{84.5 \times 143}{84.5 - 143} = \frac{12{,}083.5}{-58.5} = -206.5$$

This image distance can now be added to the lens separation (206.5 + 130 = 336.5), and that value can be used as the object distance for the front thin lens. We can then compute the distance to its image using the same formula:

$$S' = \frac{S \times f}{S - f} = \frac{336.5 \times 286}{336.5 - 286} = \frac{96{,}239}{50.5} = 1906 \text{ mm}$$

While not the exact value of 2000 mm, this result is acceptable, and a 13-mm lens travel would be recommended (see Fig. 4-12c).

It might be noted, perhaps with some concern, that this result (object distance = 1906 mm) is nearly 5 percent off from the target value. This greatly exceeds the previously referenced <1 percent rule of thumb. It is always important to keep in mind the real meaning of the numbers that are being generated. In this case the goal is to be able to focus the lens down to objects as close as 2000 mm. Accepting a solution that reached only 2100 mm would be an error which might result in an unhappy customer. On the other hand, the solution we have found, which can be focused down to 1900 mm, is not a problem. In fact, it will allow for manufacturing and assembly tolerances and still meet the 2000-mm minimum focus goal.

It will be recalled that in an earlier example, it was concluded that a possible overall cost savings would result due to the elimination of a focus mechanism requirement. In the case of this second example, the requested change does not seem to offer a similar cost advantage. Why then, one might ask, would this approach, of moving just the rear element rather than the entire lens, be considered advantageous? That would be a valid question and it is one worth considering in more detail at this point.

In many designs the internal volume of a precision lens assembly will be purged and filled with pressurized dry nitrogen to prevent the accumulation of dirt and/or moisture on internal lens surfaces. If that were the case for this lens, and our design called for the detector to be moved away from the lens to achieve focus, there would be a substantial change in the interior volume of the lens assembly as the focus was adjusted from infinity to 2000 mm. As this increase in volume took place, the tendency would be for outside air to be drawn into the lens, along with any dirt or moisture that might be contained by that air. This would obviously defeat the purpose of having a purged assembly.

On the other hand, our second approach, which involves moving just the rear lens element, offers the obvious advantage that the internal volume remains constant throughout the entire range of focus travel (see Fig. 4-13). The only precaution required here would be that a passageway must be provided as shown, to permit the dry nitrogen to flow between the forward and rear volumes of the lens assembly as focus is taking place. It is through the understanding of basic design principles such as this that the optical engineer is able to better appreciate the importance of his work and, as a result, able to do a better job at it.

Whenever a design judgment such as this is implemented, it is important to consider what tradeoffs might be involved. In everyday terminology, you hardly ever get something for nothing. In this case, we have gained the advantage of an assembly where the internal volume is constant throughout focus. In order to accomplish that, the separation between the two thin lenses has been varied by a substantial amount. It will be recalled from the thin-lens formula shown in Fig. 4-5 that a change in lens separation will produce a change in EFL. It should be considered whether the amount of change involved here will have a negative impact on overall system performance. We can calculate the lens EFL when the focus is set at the minimum range of

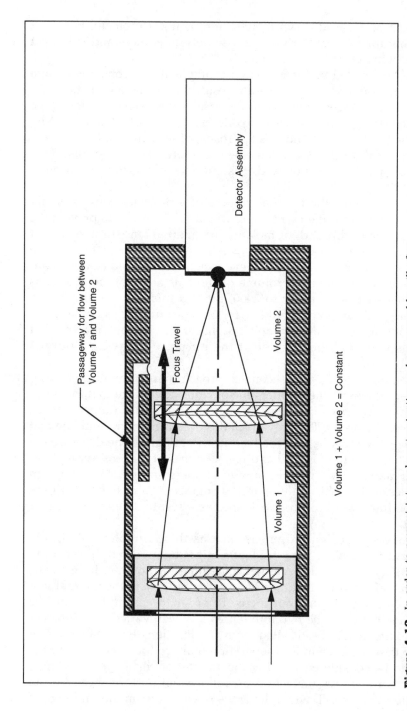

Detector Assembly

Passageway for flow between
Volume 1 and Volume 2

Focus Travel

Volume 2

Volume 1

Volume 1 + Volume 2 = Constant

Figure 4-13. In order to prevent internal contamination, a lens assembly will often be sealed, purged, and filled with dry nitrogen under a slight internal pressure. The concept of moving the rear lens element for focus maintains the internal volume constant.

2000 mm. For that case, the lens separation was reduced from 143 to 130 mm. The resulting thin-lens EFL will in turn be reduced from 143 down to 137 mm, a decrease of about 4 percent. While this is not a negligible amount, in most cases it would probably be deemed a reasonable tradeoff, considering the design advantage that has been gained.

4.11 Mock-Up of the Thin-Lens System

There is an intermediate step, between thin-lens theory and final thick-lens design, that often permits valuable information to be generated at very little cost in terms of time and money. I refer to this as the *simple lens mock-up stage*. We have stated that the thin lens is not a reality, it is only a concept. In terms of actual hardware, the simple lens is the nearest thing to a thin lens. This is a single lens element, constructed from standard optical glass. The simple lens will have the diameter and the optical power of the thin lens, while still not dealing with the fundamental problem of image quality. There are two ways in which the optical system can be carried from the thin lens to the simple lens mock-up stage. First, it can be set up for computer analysis using one of several optical design software packages available today. Alternately, it is possible to purchase simple lenses from several sources and to physically mock up the system on a simple lens bench. In this section we will concentrate on the computerized approach, since it offers a great deal more flexibility and is more in keeping with recent developments.

The computerized simple lens mock-up requires access to a basic computer system, including hardware and optical design software. While the experienced lens designer would traditionally have access to such a system, it is becoming increasingly common, and beneficial, to provide similar tools to all optical engineers and systems designers. It is obviously quite important that these personnel have the opportunity to become familiar with the basic steps involved in the use of such programs. This is a relatively new concept, but an important one that should be given serious consideration. While the general optical or systems engineers are not (yet) expected to generate final lens designs, the availability of modern computer systems is tending to expand their responsibilities into the area of preliminary lens design and analysis. While there are a number of suitable computer systems available, the examples presented in this chapter will assume a specific con-figuration, including a 486/33 PC, on which the OSLO (Optical System and Lens Optimization) Series 2 optical design package, available from Sinclair Optics, has been installed.

For this exercise, let's consider once again the Petzval lens form that is shown in Fig. 4-5. In order to simulate this lens using the computer system, we will first need to assign some basic physical characteristics to the two thin lenses being simulated. One bit of advice that I was exposed to many years ago is worth repeating at this point. In optical system design and analysis a picture of the lens, as it is seen by the computer, is worth considerably more than a thousand words. The primary reason for this is that, in doing mathematical ray tracing, the computer does not generally consider the physical characteristics of the optical components. For example, a lens may have a negative edge or center thickness and the computer program will still happily trace rays through that lens. By drawing a picture of those optics, or better yet by having the computer generate a graphic, it becomes readily apparent when an unacceptable physical condition has been assumed or generated.

For this example, we want to produce a computer model made up of two lenses that will simulate real, simple lenses that might be pur-chased from an optics supplier to mock up our thin Petzval lens. We know that the first lens needs to have a diameter of about 75 mm and a focal length of 286 mm. With a separation of about 143 mm, we then need a second simple lens that is about the same diameter, with a focal length of 143 mm. Assuming both lenses to be planoconvex, we can estimate the first radius using the simple lens approximations that are given in Fig. 4-14. From these data we conclude that the first radius of the first lens will be equal to one half its focal length, or in this case 143 mm. The thickness of the element may be determined by a basic rule of thumb that states the lens thickness should be at least one eighth the lens diameter. Knowing this, we can assign a thickness of 8 mm to both lenses. The material for both lenses will be input to the computer as BK7-type crown glass. To compensate for the 8-mm lens thicknesses, we will reduce the 143-mm separation between lenses to 135 mm.

All optical design computer programs include the ability to assign a value to the angle of the marginal ray as it leaves a surface, in place of a radius on that surface. The program will then automatically adjust that radius to produce the exit angle that has been requested. This very useful feature is known as an *angle solve*. From earlier calculations we know that the height of the marginal ray at the first lens is 35.75 mm. We also know that this lens has a focal length of 286 mm. Thus,

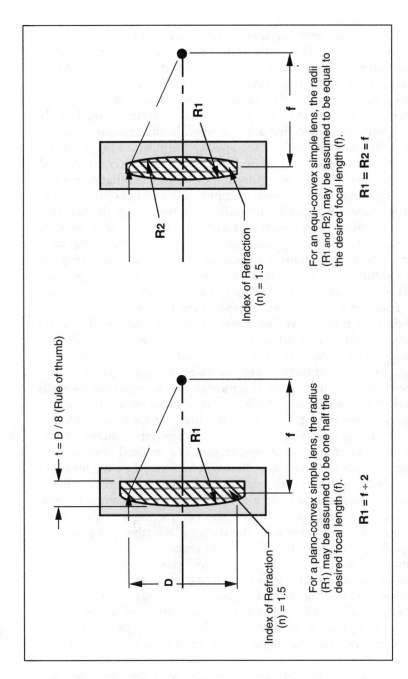

Figure 4-14. The thin lens can be simulated by a simple lens, for purposes of preliminary computer analysis. The above approximations show the relationship between lens focal length *f* and the radii of the two most common forms of simple crown glass lens elements.

the angle of the marginal ray leaving this lens will be 35.75 ÷ 286 = − 0.125. We can assign this value as an *angle solve* on the back surface of the first simple lens. Using the same approach for the second lens, we can conclude that the first surface radius is 71.5 mm, while the second surface has an angle-solve value of − 0.250. The thickness from the last lens surface to the image plane can also be determined by the computer, using what is called a *height solve.* In this case, instead of a thickness value at that surface, we specify the desired height of the marginal ray on the next surface (in this case, zero). The computer will then determine the thickness required in order to meet that constraint. The resulting simple lens system, as input to the OSLO optical design program, is shown in Fig. 4-15. In addition to the lenses, the marginal and chief rays have been traced and are also shown in that figure. It can be seen that our assumption that the second lens be of the same diameter as the first (75 mm) has resulted in an unacceptable physical condition at the edge of the lens in the form of a negative edge thickness. This is quite easily remedied by reducing the diameter of the second lens from 75 to 45 mm and accepting the clipping that occurs for the off-axis light bundle. At this diameter, the rear lens will pass the entire axial bundle. The off-axis energy loss due to that clipping (vignetting) is estimated to be about 25 percent, a perfectly normal and acceptable amount of vignetting at the maximum field angle.

This percent vignetting figure is typical of the data that can be easily confirmed by ray tracing the simple lens configuration on the computer. In this case, the appropriate clear apertures are assigned to the two lenses, followed by spot diagram calculations from the center and then the edge of the field of view. A *spot diagram* is a method of simulating the image of a point source object that will be formed by the lens. The entrance pupil of the lens is divided into an arbitrary large number of equal areas (80 in this case) and a ray is traced from a common object point through the center of each area. By constructing a graphic of the intersection of those 80 rays with the image plane (the spot diagram), the computer produces a fairly accurate representation of what the actual image of a point source would look like. When vignetting is present, some of these 80 rays will not pass through the optics. The computer program will indicate the percentage of rays that have been blocked, which will yield a very close approximation of the amount of actual vignetting that is present in the lens system. In this case the computer confirms that the vignetting at the maximum field angle is 30 percent.

With this model of the thin-lens system now in the computer, it is possible to revisit the focus problem that was covered earlier using thin-lens formulas. The angle solves that were used to determine the

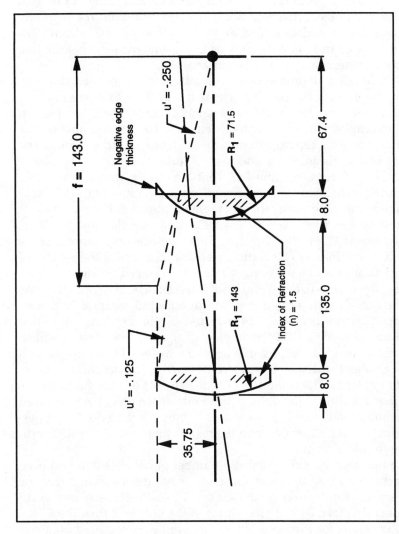

Figure 4-15. The 143-mm, $f/2.0$ Petzval lens, developed using thin-lens theory, has here been modified to simple lens form for computer analysis. The simple lens approximations in Fig. 4-14 have been used to establish the R_1 radii. Angle solves have been assigned to establish the radii of the back surfaces.

second radius on each lens have served their purpose, and must be removed prior to changing the object distance. The height solve, on the other hand, will be left in place so that the computer will automatically recompute the new distance to the image plane each time a change is introduced to the lens system. Within the computer model, we can now move the object from infinity, to a point 2000 mm from the lens. With the object at infinity, the image distance was found to be 67.4 mm. When the object distance is changed to 2000 mm, the resulting new image distance (determined by the height solve) is 77.6 mm. This 10.2 mm change in image distance is consistent with the results that were found using our earlier thin-lens analysis.

We can now simulate the case where the lens is focused down to 2000 mm by moving the rear lens element only. Our thin-lens analysis indicated that a movement of the second element 13-mm to the left would accomplish the desired focus change. To simulate this on the computer, the lens separation is reduced from 135 to 122 mm. The resulting image distance is found to be 80 mm. The goal in this case is to maintain the distance from the front lens to the image plane constant during focus. In order to accomplish this, the sum of the lens separation plus the image distance must remain constant. For the object at infinity that sum was $135 + 67.4 = 202.4$ mm. For the object at 2000 mm it becomes: $122 + 80 = 202$ mm. This confirms once more that the results of the thin-lens calculation were correct. Using the results of the final thin-lens calculation, we can now fine-tune the focus by changing the object distance and noting the change in image distance. It can be quickly determined that, for an object distance of 1900 mm, the image distance becomes 80.4 mm and the sum of lens separation plus image distance is 202.4 mm, exactly equal to the starting value. This exercise tells us that, by moving the rear lens 13 mm to the left, we will change the distance at which the lens is focused from infinity down to 1900 mm. The computer output also tells us that, for this refocused lens condition, the EFL of the Petzval lens has been reduced from its initial value of 143 down to 137 mm. Once again, this 4 percent reduction in EFL agrees with the results that were generated earlier using thin-lens formulas.

While this exercise has revealed a number of answers that had been easily achieved using thin-lens formulas, it has demonstrated that the use of a simple computer model makes it possible to generate additional data that take us one step closer to the ultimate, final thick-lens design. We were, for example, able to determine the required diameter of the second lens element and to compute the amount of vignetting that would result and to confirm results found earlier dealing with the lens focus problem. It can be seen that any number of other system parameter changes could easily be entered into the computer model (e.g., wavelength, f number, field of view) and the results evaluated instantaneously. The computer and associated software should be regarded as an extension of the optical engineer's arsenal of tools, not just for detailed lens design and analysis, but also for several aspects of preliminary lens design and system calculations.

4.12   Review and Summary

The thin-lens concept represents a method by which the lens designer or optical engineer can represent the final optical system, without having to deal with the complications of a thick-lens configuration. Using the thin-lens approach, it is possible to establish all the basic lens system parameters by applying simple ray trace methods and formulas. Early thin-lens analysis can be used to determine the relationship of the aperture stop, the entrance pupil and the exit pupil, to the lens system. The question of image quality must be reserved for later, thick-lens design and optimization.

Each thin lens has a degree of optical power, which is found by calculating the reciprocal of its focal length. Formulas and approaches have been presented, demonstrating how the power of several thin lenses can be combined. It has also been shown how the tracing (mathematically or graphically) of a few basic rays through a thin-lens system will reveal significant information, especially in terms of lens diameter, power, angular field of view, image size, and location. In lieu of ray tracing, it is possible to conduct similar thin-lens analysis using the gaussian or newtonian thin-lens formulas to establish object–image relationships. A number of typical examples have been covered, illustrating the application of thin-lens theory to the solution of actual problems.

Finally, we have devoted the final section of this chapter to a discussion of how a modern-day PC system might be used by the optical engineer to execute thin-lens system analysis. This is felt to be a very important concept, representing a revolutionary method for the optical engineer to apply readily available computing tools to the early stages of optical system analysis. Once the engineer has created a simple lens mock-up of the optical system using the computer, it is much the same as having an actual lens system set up on a lens bench in the optics laboratory, where any number of variables may now be conveniently introduced to the system and resulting changes in related parameters can be observed.

In the next chapter the reader will be introduced to an all-new software package, designed specifically for use by the optical engineer in this type of preliminary system design and evaluation. This new computer program will eliminate the need for the engineer to learn the details of a more complex, full-featured lens design program. In addition, this new software package facilitates the inclusion of a wide variety of commercially available optical components, making possible an even more effective simulation of actual optical hardware within the computer environment.

5

Optical Design with OSLO MG

5.1 Introduction

This chapter will introduce the reader to another aspect of today's computer-based methods of optical engineering and lens design. Until recently, the topic of lens design would have been treated as a very specialized, almost mysterious subject, and it would have been confined to an isolated chapter of any book dealing with optical engineering. This chapter will introduce the concept of the *optical designer*, this being the individual responsible for executing that combination of tasks previously accomplished individually by either the *optical* engineer or the lens *designer*.

There is a definite trend today toward an environment where the area of expertise to be covered by the optical engineer and the lens designer has been broadened to the point where considerable overlap now exists. The person doing the combined work of the optical engineer and the lens designer can reasonably be referred to as an *optical designer*. This chapter will describe the work of the optical designer and demonstrate some of the recent technological developments that make this new position possible, particularly in the areas of computer hardware and optical design software.

5.2 A Historical Perspective

This entire discussion will become a bit more meaningful if a real timeframe is introduced. The publication date of this book is 1994...I have worked in the field of optical engineering for about 30 years...when I refer to *recent* developments, this would involve the decade from 1985 to 1994.

Owing to the very unique nature of the work involved, the lens designer has in the past worked alone (or with a group of other dedicated lens designers) in generating and analyzing new lens designs. Once completed, the designer's work would typically be turned over to the *optical* or *systems* engineer for further analysis and incorporation into the final product. If, at any point, a lens-design-related question were to come up, it would be referred back to the lens designer for resolution. In other words, the lens designer had very little to do with the overall design of the system, while the optical engineer had nothing to do with lens design. This all worked reasonably well, but it did result in a high degree of specialization, especially in the case of the lens designer.

More recently, revolutionary developments have taken place in several areas that have led to a breakdown in this specialization and a blending of many functions that are now shared by the optical engineer and the lens designer. While this concept is impossible to quantify precisely, the illustration in Fig. 5-1 reflects graphically the general nature of these developments. With the exception of a few isolated areas that remain the exclusive domain of one or the other, the lens designer is now called on to consider a variety of systems-related aspects of the design, while the optical engineer is expected to handle some day-to-day lens design tasks, particularly in the areas of image quality, tolerance, and environmental analysis.

The two major technological changes that have made this blending of responsibilities possible are the development and growth of the personal computer (PC) and the availability of optical design software that can be run on the PC. A historic review (from the perspective of this author) will be helpful in demonstrating the magnitude and significance of these developments. My experience as a lens designer began with a major aerospace corporation in 1962. This operation involved the lease of an IBM-1620 computer, along with an integrated optical design software package. The hardware consisted of a computer with a typewriter-style printer and a card reader-punch. This was a fairly compact installation for its time, requiring a dedicated, air-conditioned computer room that was about 15 × 25 ft in size. Lens optimization was accomplished by controlling third-order aberrations. Analysis of image quality was done by calculating aberration data and spot diagrams. Any plotting, mostly aberration curves and lens drawings, was done manually by the lens designer at the drafting board. In terms of cost, this system was leased for about $2500 per month, which, as a

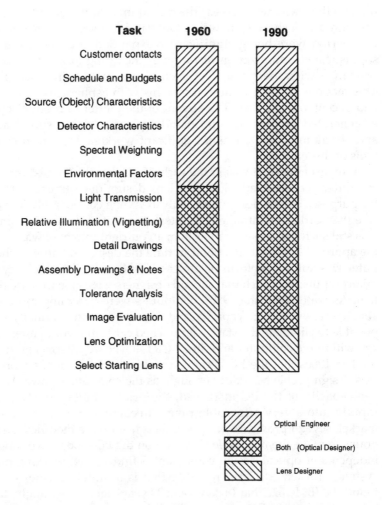

Figure 5-1. Until recently, the typical tasks of an optical design project have been divided quite clearly between the optical engineer and the lens designer. Today, with the advent of the personal computer and modern optical design software packages, the overlap between the two disciplines has increased dramatically.

point of reference, was about three and one half times my monthly salary as a junior engineer at that time. Later in the decade of the 1960s the large mainframe computer became the machine of choice for lens design. Of course, the cost of such a machine was astronomical, making access a problem. One could travel to a large computer center and rent time on the machine, or access the mainframe computer via phone lines through a timesharing network. Optical design software packages of this period were being significantly expanded and improved. More sophisticated optimization techniques, including exact ray trace data combined with aberration coefficients, were now being incorporated. This made it possible to design systems with aspheric as well as tilted and decentered surfaces. The advent of the plotter to produce computer generated graphic output, including lens pictures, spot diagrams, and aberration curves, was a significant work saving addition to systems of this era.

In my experience, it was in the early 1970s that the "mini-" mainframe computer became available to the lens design department, exclusively for purposes of lens design and analysis. While considerably less expensive than earlier mainframe computers, the cost of these minicomputers was still such that leasing rather than an outright purchase was generally the approach used. At this point in time the designers input to the computer was still in the form of punched *IBM cards*. The efficiency of the minicomputer was such that several designers were able to share the system. As we entered the 1980s, computers were becoming more affordable. Increased storage capacity and improved methods finally made it possible to eliminate the stacks of punched cards in favor of internal storage within the computer and direct CRT–keyboard interfacing. It was in the late 1980s that the PC became a viable platform for the working of lens design problems. Oddly enough, as the computing power and storage capacity of the PC increased, the price simultaneously dropped rapidly into a very affordable range. Recognizing the great potential market, several providers of optical design software modified their programs to make them suitable to be run on the PC. Today, working as an independent optical design consultant, I find myself working on a PC system with lens design capability that is infinitely greater than that found the IBM-1620 that I used some 30 years ago. Interestingly, the cost to purchase this PC system outright is just about equal to the $2500/month lease payment of that era.

This is more than your typical "walked 6 miles to school" old-timer's story of how tough things used to be. For example, while (like the computer) the automobile that we drive today is considerably more sophisticated than those produced in 1962, its cost has increased by more than 10 times during that same time period. In the case of the computer in general, and optical design systems in particular, we have reached a point where the performance has increased by many orders of magnitude, while the cost has come down to the point of absolute affordability.

5.3 The OSLO Optical Design Package

On the subject of software for optical design, I would like to briefly review the history of one specific product. This is not the only product of this type, it is not the most powerful, nor is it the least expensive. It is however, a fine product, one that is representative of the present state of the art and one with a history that demonstrates many of the more significant developments in this field during these years. It is the software package that will be used to generate the design examples that follow in this chapter as well as those found elsewhere throughout this book.

OSLO (Optical System and Lens Optimization) is an optical design software package that has been evolving for nearly 20 years. A product of Sinclair Optics, OSLO has become one of the best known and widely used optical design programs in the world today. Doug Sinclair, president and founder of Sinclair Optics, received his doctorate from the University of Rochester in 1965. After several years as a professor at the Institute of Optics, he established Sinclair Optics as a part-time venture in 1976. The company charter has been to provide state-of-the-art optical design software to a rapidly growing group of users. Initially a very basic program, run on early versions of the desktop computer, OSLO has developed over the years to a position where it challenges for the position of number one in the industry.

Recognizing contemporary trends, OSLO was one of the first professional-level optical design programs capable of being run on the PC. Most recently, the entire line of OSLO programs has been upgraded to run using the Microsoft Windows interface on the PC. This combination (OSLO, the PC and Windows) results in a program with total capability, combined with a previously unheard of degree of user-friendliness. This has precipitated the situation described earlier, where much of the mystery that had been associated with the lens design and analysis process has been swept away, making it possible for the optical systems engineer, and many others working in closely related areas, to perform a wide variety of basic optical design tasks. At the same time, this has freed lens designers of many tedious tasks, allowing them to greatly broaden the scope of the work in which they are involved. Certain aspects of tolerance analysis, environmental analysis, and numerous other areas of more general optical engineering and system design have now been shifted to a position of shared responsibility between the optical engineer and the lens designer.

5.4   Introduction to OSLO MG

The lens design examples presented in the last chapter were executed using the Series 2 version of OSLO. It might reasonably be expected that most engineering organizations seriously involved in optical design would have access to Series 2, or an equivalent optical design software package. The examples to be covered in this chapter will utilize the OSLO MG optical modeling software package as the basic design tool for the solution of a series of similar problems. This is the most basic version of several forms of OSLO available today. It has been priced and marketed in such a way that it is available to a much larger audience than are most other optical design software packages. In this case it is assumed that the problem solver is not an experienced lens designer but rather, one who is closely associated with the areas of preliminary system design and specification. The following examples will demonstrate the ease of use and the effectiveness of the OSLO MG package, which is typical of several low-cost programs that are readily available to today's optical designer.

OSLO MG runs on a modern PC utilizing the Microsoft Windows interface. When the program is active, several windows will be seen on the monitor simultaneously. The *main window* (see Fig. 5-2) is the central control point for all design tasks. It contains several graphical components for running the program, and it is always active.

Key components of the main window identified in Fig. 5-2 are

Title bar	Identifies the lens being worked on
Menu bar	Shows the top level items of the OSLO menu system
Tool bar	Contains several buttons with icons indicating common tasks
OSLO control and mode buttons	For program operation
Prompt line	Contains a prompt indicating program status or requesting input
Command line	Keyboard entry line for input to OSLO

Other windows may also be present, depending on the program function that is active at that time. Most common windows are the text window, graphic window, catalog lens database window, and the data editing spreadsheet windows. Figure 5-3 illustrates a typical screen condition with several windows present.

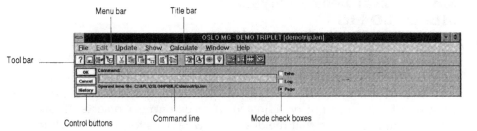

Tool bar

Menu bar   Title bar

Control buttons   Command line   Mode check boxes

Figure 5-2. The main window of OSLO MG is the central control point for all tasks. It contains several graphical components and it is always active.

Documentation provided with the program is quite comprehensive. This chapter is not intended to duplicate the function of that documentation package, but will demonstrate several of the programs more significant features by applying it to the solution of some practical design problems.

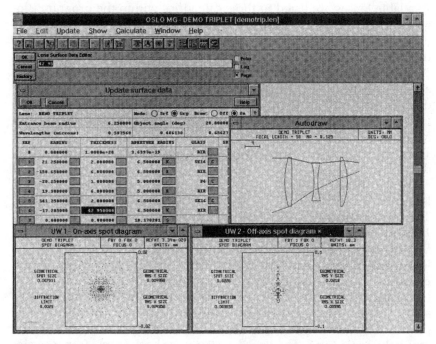

Figure 5-3. While running OSLO MG, several windows will be simultaneously active. This illustration is a typical screen condition.

5.5 Petzval Lens Analysis with OSLO MG

In several examples presented in previous chapters a Petzval lens with a focal length of 143 mm, a 10° field of view, and a speed of $f/2$ was considered. This section will demonstrate how OSLO MG might be used to perform the preliminary analysis of this lens, including the question of how it might be adjusted to focus on near objects. It has been determined earlier that the desired Petzval lens form will result if we use a two-lens configuration, where the first lens has a focal length of 286 mm and a diameter of about 75 mm. The second lens, spaced about 143 mm from the first, will have a focal length of 143 mm and a diameter of about 50 mm. The OSLO MG program can be used to simulate this system for computer analysis.

The first step in this process is to enter the catalog lens selection mode and search the catalog database for a pair of lenses with parameters that are close to those just outlined. This database contains a series of optical components that will be found in the inventory of Melles Griot, an established supplier of optical products for many years. Available lenses are listed in three categories within the program: singlets, doublets, and other. The lens selection limits are set by the designer, based on EFL range, lens diameter range and lens type. For the first element, an EFL of 286 ± 20 and a diameter of 75 ± 10 is entered. A search of the singlet category reveals that no suitable lens is available. Switching to the doublet category, it is found that a 300-mm doublet with a diameter of 82 mm is available. This lens has a reference catalog number of LAO267. While the EFL of this lens is a bit longer than our target, this lens will work for the intended purpose. The doublet lens (LAO267) is copied from the catalog into the lens data spreadsheet and the next step begun.

For the second lens, the *merge catalog* mode is used. Here a second lens from the catalog can be selected and added to the existing system. The singlet category is searched for a lens with a focal length of 143 ± 10 and a diameter of 50 ± 5. A viable candidate lens is found with an EFL of 150 and a diameter of 50 mm (catalog number LPX241). This lens is added to the spreadsheet following the doublet that was selected earlier. To simulate the thin-lens separation of 143 mm, these two lenses are set at an arbitrary distance of 130 mm. The paraxial lens setup data are then modified by calling up a second spreadsheet-like display. Using this spreadsheet, the lens f number is set to $f/2$ and the semifield angle to 5°. The image distance, or back focus, from the singlet to the image plane is determined by assigning a paraxial height solve to the last lens surface.

Examining the paraxial data for this lens system, we find that the EFL is 145.2 mm. This was reasonably close (within 2 percent) to our goal of 143 mm, but can be easily fine tuned by adjusting the lens spacing which had arbitrarily been set to 130 mm. A couple of trials show that a lens spacing of 125 mm will result in the desired 143-mm EFL for the lens combination. The lens data and paraxial setup are then printed out, the results are shown in Fig. 5-4. The graphic mode of OSLO MG can be used to generate the standard lens layouts shown in Fig. 5-5 and the solid model version seen in Fig. 5-6. The advantage of the Microsoft Windows interface is obvious in all the work done up to this point. The spreadsheet format makes changes to the lens system both simple and intuitive. Pull-down menus and click-on buttons allow the quick selection of complex analytical tools, without any knowledge (or memory) of complex commands and sequences of input data. The text and graphics output is available, generally to a high-quality laser printer, with a simple click of a mouse button.

```
*LENS DATA
SIMULATED PETZVAL
  SRF      RADIUS       THICKNESS    APERTURE RADIUS     GLASS   SPE    NOTE
   0        --         1.0000e+20    8.7489e+18           AIR

   1        --            --         35.750000 A          AIR

   2      ELEMENT  F    19.600000 F  41.000000 F         FIXED F  LAO267
   4      LAO267   F   125.000000    41.000000 F          AIR     LAO267

   5        --            --         22.500000 K          AIR

   6      ELEMENT  F     7.400000 F  26.000000 F         FIXED F  LPX241
   7      LPX241   F    73.982968 S  26.000000 F          AIR     LPX241

   8        --            --         12.501001 S

*PARAXIAL SETUP OF LENS
  APERTURE
    Entrance beam radius:        35.750000    Image axial ray slope:      -0.250198
    Object num. aperture:      3.5750e-19     F-number:                    1.998421
    Image num. aperture:          0.250198    Working F-number:            1.998421
  FIELD
    Object angle:                 5.000000    Object height:             -8.7489e+18
    Image height:                12.501001
  CONJUGATES
    Object distance:           1.0000e+20     Object to prin. pt. 1:     1.0000e+20
    Image distance:              73.982968    Prin. point 2 to image:     142.887094
    Overall lens length:        152.000000    Total track length:        1.0000e+20
    Paraxial magnification:    -1.4289e-18    Effective focal length:     142.887094
  PUPIL DATA
    Entrance pupil radius:       35.750000    Entr. pupil position:          --
    Exit pupil radius:          412.625042    Exit pupil position:       -1.5752e+03
```

Figure 5-4. Content of a typical text window in the OSLO MG program. Shown are the lens data and the paraxial setup for the simulated 143-mm, $f/2$ Petzval lens constructed using two components from the catalog database.

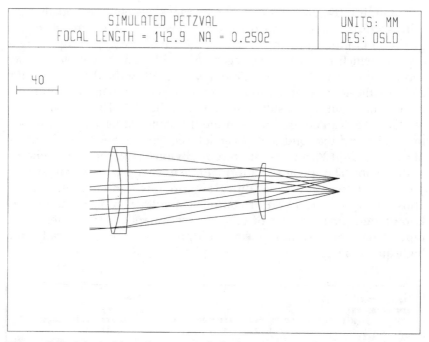

Figure 5-5. A typical lens drawing, including ten rays, as generated by the OSLO MG system design program.

While it is not expected that this model will reveal any useful information relating to image quality, we do now have a computer simulation of the Petzval lens on which to conduct a variety of analyses that will accurately predict the behavior of the actual lens in several areas. This step in the design process is intended to provide more general information that will be used by the lens designer to generate the final design at a later point. For example, thin-lens analysis done in Chap. 4 indicated that the diameter of the second lens group could be reduced to 45 mm without introducing an unacceptable amount of vignetting. Using this computer model, it is a simple matter to add dummy surfaces just before the first and second lens groups, and to assign clear apertures of 75 and 45 mm to those surfaces. The lens data is set to the mode where these apertures are checked to determine the passage of rays. It is now a simple matter to click on the off-axis spot diagram button and the information appearing in the text window tells us that the off-axis transmission (relative to the on-axis) is 69 percent. This indicates a perfectly acceptable 31 percent vignetting value at the maximum field angle of 5°.

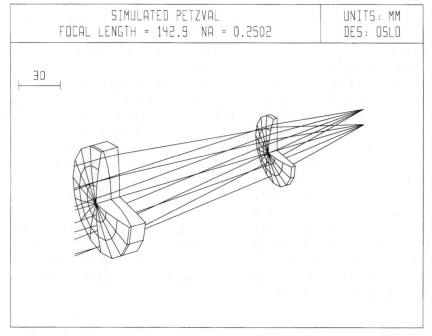

```
  SIMULATED PETZVAL                          UNITS: MM
FOCAL LENGTH = 142.9  NA = 0.2502            DES: OSLO
```

30

Figure 5-6. A solid model (3D) lens drawing, including 10 rays, as generated by the OSLO MG system design program.

We are also interested in the particular question of how this lens might be adjusted to focus on objects as close as 2000 mm. While we have answered those questions quite adequately earlier, using thin-lens formulas and then using the Series 2 lens design program, we can now consider these same questions using OSLO MG, a tool that is not only much simpler to use but is also available to a much wider and more diverse group of potential users.

Calling up the lens surface data update spreadsheet, it can be seen that the back focal distance for an object at infinity is 74 mm. It will be recalled that this back-focus dimension has been found by using a height solve on the last lens surface. By simply changing the number in the object distance box of the spreadsheet to 2000 mm, it will be seen instantly that the back focus changes to 84.2 mm. This 10.2-mm focus shift agrees with the value that was found earlier by two different methods.

The alternate focus approach that we want to considered involves axial adjustment of the rear lens while maintaining the overall length of the lens assembly constant. Rather than the paraxial height solve that has been used, it is possible to set the back focus such that the sum of its length plus the dimension between lenses remains constant. In this mode, any change to the space between lenses is applied with an opposite sign to the back focus. This constraint is added with the object distance set to infinity. Using the spreadsheet mode to change lens surface data, the space between the lenses is designated as a variable and the object distance is changed to 2000 mm. Because the lens focus had been set for infinity, the lens system is now out of focus by a considerable amount. Using the autospace feature, the distance between lenses can now be varied automatically to bring the system back into focus. The new space between lenses is found to be 112.8 mm, a shift of 12.2 mm from the initial value. Meanwhile, the constraint that was placed on the back focus maintains the overall length of the system as a constant. It had been concluded from earlier analysis that a lens travel of 13 mm would bring the point of focus in to 1875 mm. The spreadsheet value for object distance can be changed to 1875 mm, the autospace repeated, and the amount of lens travel will be found to be 13.0 mm, confirming once again the results that had been found by previous methods.

5.6 Laser-Transmitting Lens System

This next example demonstrates methods that might be employed in using OSLO MG as a tool to select a commercially available optic for the transmission of laser energy from one optical fiber to another, with a 100-m propagation path between fibers. This example has appeared in several forms of documentation prepared for the demonstration of OSLO MG's basic features and capabilities. It will be reviewed here and expanded on to some degree.

The general system approach here will be to collect the laser energy output from the first fiber, collimate that energy, and project it over the 100-m distance. At that point an identical lens will be used to collect the energy and focus it down onto the second fiber. Figure 5-7 illustrates this system, along with certain physical constraints that have been imposed on the lens assembly. Since the fiber diameter in this case is 5 μm, it is reasonable to assume that the spot size of the chosen lens must be substantially less than that 5-μm dimension. The optical designer's task here is to find a commercially available lens that will meet all requirements of this application. While such a selection process might have been accomplished in the past by actual lens bench testing in the

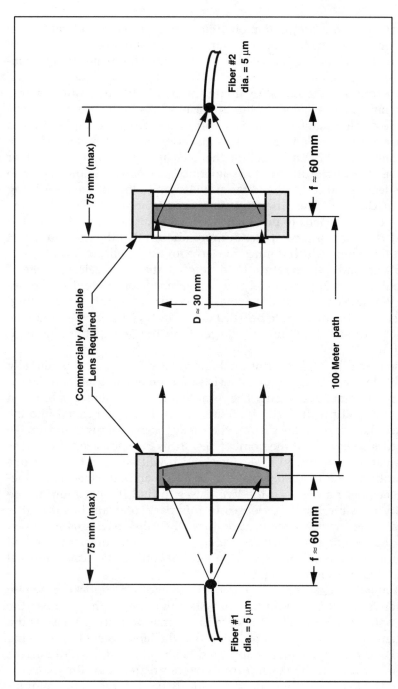

Figure 5-7. The optical design task here is to select a commercially available lens assembly suitable for the transfer of laser energy from optical fiber 1 to optical fiber 2.

laboratory, this example demonstrates how a modern optical design program permits us to effectively simulate that laboratory testing on the computer. In so doing, the savings in both time and money will be substantial.

As a first try, we might evaluate the possibility of using a single planoconvex lens element in this application. As in the previous example, the catalog database that is a part of the program can be used to locate and select a planoconvex lens that is 30 mm in diameter and has a focal length of 60.223 mm. Calling that lens into the program for further analysis, the Microsoft Windows interface makes it a simple matter to set its lens speed to $f/2.0$ and the wavelength to that of the laser being used (0.6328 µm). The program is then used to automatically adjust the back focus of the lens to that point where the focused spot size is minimized. At that point, computer output indicates that the minimum spot size is 0.195 mm, which is much too large for this application.

For a second try, we return to the lens database and select a cemented doublet with approximately the same dimensions to replace the singlet. We quickly repeat the above procedure and find the resulting spot size for the doublet to be 0.022 mm, which is much better than the singlet, but still substantially larger than the 5-µm fiber size with which we are dealing.

Reviewing the Melles Griot catalog, we see that there is a potentially suitable lens combination that consists of a cemented doublet followed by a positive-meniscus lens. The focal length of this lens combination is 60 mm, and the diameter is 30 mm. These lenses are called into the surface data spreadsheet using the *catalog merge* feature, and image quality is analyzed. At the point of best focus the spot size for this lens combination is found to be 1.8 µm, which will be acceptable for this application. One final step that can be easily accomplished is to fine-tune the spacing between the two lenses to find the optimum image quality possible for these lenses in this particular application. It is found using the *autospace* feature, that when the spacing is increased from the recommended catalog value of 1 to 5 mm, the spot size will be reduced from 1.8 down to 1.3 µm. A layout of the resulting final lens system is shown in Fig. 5-8.

A follow-up example will demonstrate another application where the convenience of computer analysis shines through. Suppose the laser optics shown in Fig. 5-8 were in the manufacturing phase and a call came from the optics lab stating that the lens assembly was being tested and the image quality appeared to be quite bad. In addition, the BFL was found to be several millimeters greater than the expected value. The lens design will have previously been stored in the comput-

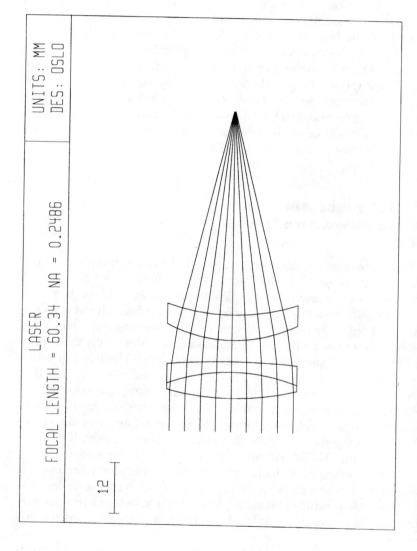

FOCAL LENGTH = 60.34 NA = 0.2486

LASER

UNITS: MM
DES: OSLO

12

Figure 5-8. A lens drawing of the two-component laser optics design generated using the OSLO MG system design program.

er's lens library for convenient recall. A quick review of the design leads to the suspicion that the doublet orientation may have been reversed at assembly. In the update surface data mode, it is possible to select the doublet and then reverse it. At this point it is the designer's responsibility to consider the effect that this lens reversal will have on the actual lens assembly. Because the first curvature is considerably steeper than the last, reversing the lens will reduce the airspace following the doublet by about 1.5 mm. All optical design work must be a product of both the computer and the designer...neither is capable of doing the complete job alone. With these changes made, the back focus is found to be increased by 2 mm and the spot size, which was less than 2 μm, is now found to be 0.09 mm. On the basis of this rapid evaluation, a likely source of the trouble has been quickly identified and the impact on lens performance accurately predicted.

5.7 Tolerance Analysis and the Interactive Design Mode

OSLO MG contains a unique new feature that adds further to its convenience and usefulness: the interactive design mode. This section will demonstrate its use as part of a typical tolerance analysis investigation. The laser optics system just designed will consist basically of a pair of commercial lenses mounted together in a simple lens cell. One of the tolerance questions that must be addressed will deal with the centration of these two lenses with respect to each other. The basic question to be answered is whether any unusual tolerances will be required to maintain the centration such that image quality remains acceptable.

When a large number of variables (tolerances) are involved, it is reasonable to establish tolerances based on statistical analysis. The basic philosophy being that it will be cost-effective to scrap a small percentage of assemblies if this will allow the tolerances to be made substantially looser. On the other hand, when small quantities are involved, it is usually best to assume a worst-case accumulation of tolerances. It is that worst-case condition that must be acceptable in terms of final performance. This second, worst-case approach, will be assumed for this analysis.

There are two tolerances that contribute to the potential decentration of a lens element. They are the outside diameter (OD) tolerance on the lens and the inside diameter (ID) tolerance on the lens cell. In this case, catalog data indicate that the dimension of the lens OD is 31.5 ± 0.15 mm. The ID of the cell must be made large enough to accommodate the maximum possible lens diameter (31.5 + 0.15 = 31.65 mm), with some additional clearance (0.10 mm) to facilitate assembly. Thus, the nominal ID of the lens cell will be 31.75. Since this is a minimum allowable dimension, the tolerance on it should be all in the + direction. The lens cell ID dimension might be given as 31.75 ± 0.10/ −0.00.

Having established these dimensions and tolerances, we can now consider the potential for lens decentration. The worst possible case would occur if both lens ODs were at their minimum allowed dimension (31.35 mm) and the lens cell ID were at its maximum (31.85 mm). If each lens was shifted by the maximum possible amount, in opposite directions, then the lenses would be decentered with respect to each other by a total amount of 0.50 mm. This condition is illustrated in Fig. 5-9. It can be simulated and evaluated quite easily using the OSLO MG computer program. The second element description is modified within the program such that it can be decentered relative to the first element. In the interactive design mode it is possible to change the amount that the second element is decentered while observing the effect on aberration curves or the spot diagram as these changes are being made. From this interactive design exercise it will be seen that, while a 0.5-mm decentration will cause the spot image to change its shape, its size does not appear to be increased dramatically. From this it can be concluded that a decentration of 0.50 mm will not seriously degrade image quality.

For a more absolute evaluation, it is a simple matter to perform a spot diagram analysis on the lens assembly, first in its nominal configuration and then another with the second lens element decentered by 0.50 mm. It will be found that the net result of this perturbation is that the RMS (root-mean-square) spot size increases from its nominal value of 1.3 μm, to a worst-case value of 1.5 μm. From this same spot diagram output we learn that the diffraction limited spot size for this lens is 1.6 μm. We conclude that the image degradation due to a 0.5-mm lens decentration is negligible in this application.

5.8 Instrument Design and Analysis

This exercise will utilize the thin-lens concepts introduced in the last chapter, combined with the computer-aided system layout, catalog

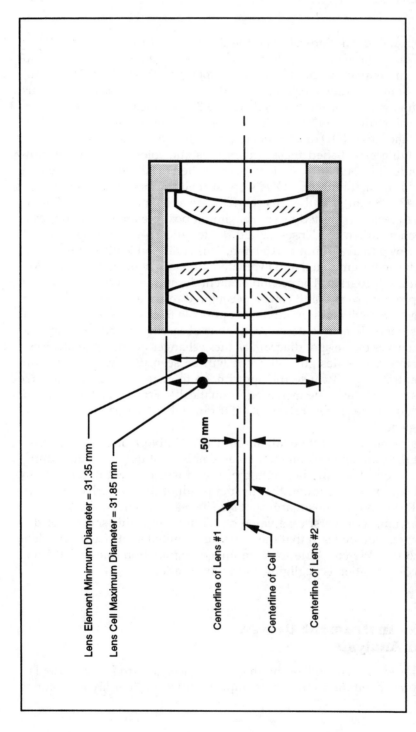

Lens Element Minimum Diameter = 31.35 mm

Lens Cell Maximum Diameter = 31.85 mm

.50 mm

Centerline of Lens #1

Centerline of Cell

Centerline of Lens #2

Figure 5-9. Lens element diameter and lens cell dimensions will determine the max-imum possible (worst-case) lens decentration within the final assembly.

lens selection, and system analysis features of the OSLO MG software package, to design a complete optical instrument.

This design will involve a low power microscope with the following specifications:

Magnification	10 ×
Object size	12 mm diameter
Working distance	>90 mm
Exit pupil diameter	3.0 mm
Eye relief	>20 mm
Image quality	Eye-limited

A microscope will typically consist of two lens assemblies, the objective lens and the eyepiece. Starting with a two element thin-lens layout (Fig. 5-10, top), one reasonable approach would be to assume that the objective lens will operate at a magnification of 1 × and the eyepiece at 10 × (EFL≈25 mm). The 90-mm working distance dictates an objective lens focal length of about 50 mm, while the overall distance from the object to the image will be approximately 200 mm.

Now, considering the actual configuration of the final system, the thin-lens arrangement can be modified such that it is made up of achromatic doublets that we might expect will be found in our lens database. Figure 5.10 (bottom) shows the system in a form that consists of four achromats, two each of two identical lenses. The eyepiece focal length will be about 25 mm. The 3-mm exit pupil diameter leads to an f number for the eyepiece of $f/8.3$. For the 1 × objective, this converts to a diameter of about 12 mm and a numerical aperture of NA = $1/(2f\#) = 0.06$, at the object. Also from the thin-lens layout, it is possible to estimate the eyepiece field of view. The image size at the eyepiece focal plane will be equal to the object, 12 mm in diameter. The half field of the 25-mm eyepiece then will be the angle whose tangent is $^6/_{25} = 13.5°$. With this preliminary lens system layout and analysis done, it is now possible to use the computer with OSLO MG to confirm the availability of the desired lenses and to evaluate the detailed optical performance of the system.

The first step is to open the catalog database and search for an achromatic doublet with a focal length of 100 mm and a diameter of about 20 mm. The closest candidate seems to be catalog number

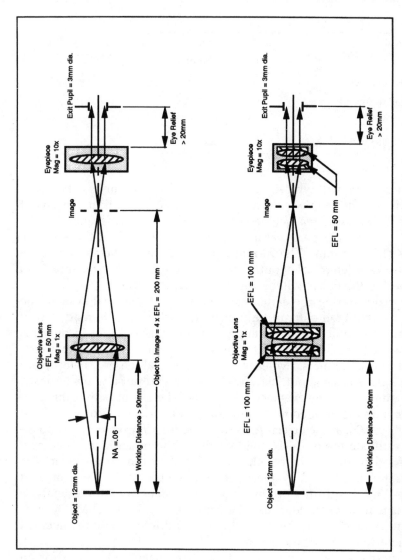

Figure 5-10. The thin-lens layout of a low-power (10 ×) microscope consists of an objective lens and an eyepiece. In the lower figure that thin-lens layout has been converted to a form made up of four achromatic doublets.

LAO123, which has the desired focal length and a diameter of 26.5 mm. This doublet is called into the lens data spreadsheet and its orientation is reversed to agree with the arrangement shown in Fig. 5-10. With this lens in the system, it is now possible to input the characteristics of the object being viewed. An object height of −6 mm, an NA of 0.06, and an object distance of 95.6 mm (a guess at this point), are entered. To control the transmitted light bundle diameter, an aperture stop will be inserted in the objective lens assembly between the two doublet lenses. Using the lens data spreadsheet, this surface, designated as the *aperture stop,* is inserted 0.5 mm from the first doublet and it is given a thickness of 0.5 mm. The second doublet, identical to the first, is then inserted using the *catalog merge* feature of the program. A paraxial height solve is assigned to the last surface of this doublet to determine the distance from that surface to the image plane. The resulting image distance is found to be 97.5 mm. If the objective lens were operating at the desired 1:1 magnification, the object and image distances would be the same. This condition can be met by changing the object distance to 96.54 mm.

The distance from the image plane to the first eyepiece element is not known, but it can be estimated to be about 20 mm. This thickness is inserted and is designated as a variable, for purposes of final system focus. The eyepiece lenses will now be added.The database is searched once again, this time for a doublet with an EFL of 50 mm and a diameter of about 20 mm. Catalog LAO059 is found to meet these requirements and is merged into the system. As with the first lens, the orientation of this lens must be reversed. A 0.5-mm airspace is assumed, and a second identical 50-mm doublet is selected and merged to complete the optical system. A dummy surface is inserted following the last element to represent the system exit pupil. Its axial location can be determined by placing a chief ray height solve of zero on the last lens surface. This will adjust the location of this dummy surface to the point where the exiting chief ray crosses the optical axis. This is, by definition, the location of the system exit pupil.

When adjusted for most comfortable viewing, the light bundles emerging from the eyepiece will be made up of nominally parallel light rays, projected to infinity. These are referred to as *collimated bundles.* A system that produces this type of collimated output is evaluated in terms of the angular error (lack of collimation) of the emerging light rays, as opposed to the actual spot size at an image plane for a conventional system. The optical design program allows a choice between *focal* and *afocal* in setting up the system parameters. In this case, the afocal mode, which evaluates angular aberrations, is selected.

The first look at system aberration curves indicates a gross out-of-focus condition. This does not come as a surprise, since the eyepiece focus position was established quite arbitrarily. It is possible to enter the interactive design mode and to adjust that thickness while observing the focus condition of the axial aberration curves. This allows us to quickly settle on the best eyepiece–image distance, which is 21.43 mm. A more direct (while perhaps less fun) method of focusing the eyepiece would be to use the autospace feature, which will automatically vary the distance from the image to the eyepiece to minimize the system aberrations. After the eyepiece focus has been set it is possible to check the distance from the last lens surface to the exit pupil plane. This is done, and the actual eye relief dimension is found to be 27.8 mm. This exceeds our target value for minimum eye relief (>20 mm) by a comfortable margin.

Examining the image quality of our system using the spot diagram routine, it is found that the axial performance is equal to about one half of the diffraction limit, while the performance at the edge of the image is about three and a half times greater than the diffraction limit. This does not represent terrible image quality, but better would be nice. Looking at the aberration curves to determine the cause of the image quality fall off with field angle, the primary problem is found to be field curvature. In a visual system such as this it will be found that the human eye is capable of compensating for a considerable amount of field curvature. This compensation can be simulated in our computer model by assigning a curvature to the object surface. In this case, trial and error reveals that a radius of 27 mm on the object surface results in a good balance of overall image quality across the field of view. That 27-mm radius represents a maximum sag at the object of about 0.67 mm. This 0.67 mm will be seen as a defocus of about 1 diopter, a perfectly acceptable value to the eye. With this amount of object curvature, the off-axis image quality is found to equal about 80 percent of the diffraction limit, a much more acceptable value. Figures 5-11 and 5-12 contain OSLO MG lens data output followed by a drawing of the final system.

If all this talk of afocal analysis and angular aberrations leaves the designer just a little uncomfortable, there is a way around it. With most programs it is quite a simple matter to select all of the optical components and reverse them all simultaneously. The object distance can then be set at infinity, the entrance pupil diameter to 3 mm and the image size to 12 mm diameter. In this configuration the rays are traced from an object at infinity, through the exit pupil, then through the eyepiece, through the objective lens, and then focused down onto the

```
*LENS DATA
10X MICROSCOPE
  SRF       RADIUS      THICKNESS   APERTURE RADIUS      GLASS   SPE   NOTE
   0      27.000000    96.500000       6.000000          AIR

   1      ELEMENT  F    6.800000 F   13.250000 F        FIXED F  LAO123
   3      LAO123   F    0.500000     13.250000 F         AIR      LAO123

   4        --          0.500000      6.000000 A         AIR

   5      ELEMENT  F    6.800000 F   13.250000 F        FIXED F  LAO123
   7      LAO123   F   96.600000     13.250000 F         AIR      LAO123

   8        --         21.409000 V    6.006590 S         AIR

   9      ELEMENT  F    5.960000 F    9.000000 F        FIXED F  LAO059
  11      LAO059   F    0.500000      9.000000 F         AIR      LAO059

  12      ELEMENT  F    5.960000 F    9.000000 F        FIXED F  LAO059
  14      LAO059   F   27.844306 S    9.000000 F         AIR      LAO059

  15        --          --            1.531827 S

*PARAXIAL SETUP OF LENS
 APERTURE
    Entrance beam radius:       5.800450    Image axial ray slope:   -2.1720e-05
    Object num. aperture:       0.060000    F-number:                  -2.196965
    Image num. aperture:      2.1681e-05    Working F-number:          2.3062e+04
 FIELD
    Object angle:               3.557852    Object height:             -6.000000
    Image height:              22.719836
 CONJUGATES
    Object distance:           96.500000    Object to prin. pt. 1:    -25.495981
    Image distance:          -7.0498e+04    Prin. point 2 to image:  -7.0557e+04
    Overall lens length:      145.029000    Total track length:       269.373306
    Paraxial magnification:    -3.786639    Effective focal length:   -25.486771
 PUPIL DATA
    Entrance pupil radius:      6.090211    Entr. pupil position:       4.820640
    Exit pupil radius:          1.531827    Exit pupil position:       27.844306
```

Figure 5-11. Content of a text window in the OSLO MG program, showing lens data and the paraxial setup for a 10 × microscope that was constructed using four optical components from the catalog database.

object surface (which has now become the image). The operating conditions must be switched back from afocal to the *focal* mode. With these changes made, spot diagram analysis shows the on axis spot size to be 0.0028 mm, which is just one half the diffraction spot size. This corresponds well (as it should) with our earlier angular evaluation. It is now possible to look at the on axis MTF curves for this system. With a 10 × microscope, the normal visual system (eye) will be able to resolve about 66 cycles per millimeter at the object. Our MTF data show the image modulation (contrast) at that frequency will be about 0.60, again a very acceptable value.

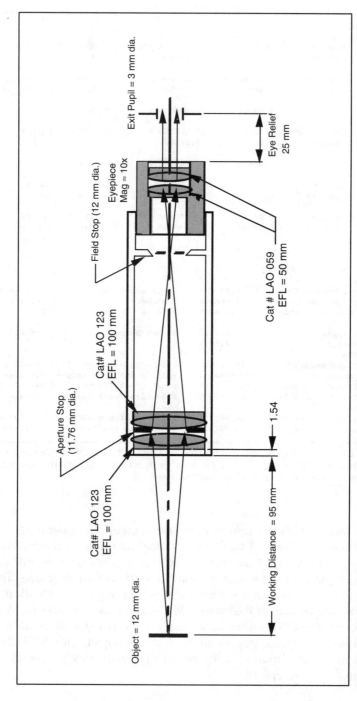

Figure 5-12. Using the OSLO MG catalog lens database, it was possible to design and analyze the system shown above. All dimensional and performance characteristics are found to meet the requirements of the original specification.

All this analysis has indicated a system that is somewhat better than the diffraction limit in all respects. It also shows that our goal of *eye-limited* image quality has been comfortably met. While this is a relatively economical solution, involving about $75 per optical component, the question will almost always come up regarding the possibility of cost reduction through simplification. The tradeoffs are easily identified and evaluated using OSLO MG. For example, one might consider substituting a planoconvex element for the lens nearest the eye in the eyepiece assembly (savings≈$50). Deleting the achromat and inserting the singlet, it will be seen that the major resulting aberration is off-axis (lateral) color, to the extent that it is now more than three times greater than the diffraction limit. A similar result is found if the field lens of the eyepiece is converted to a singlet. While possibly an acceptable solution, the performance tradeoff vs. cost savings does not appear to be acceptable in the judgment of this designer.

5.9 Magnification Analysis

This is a convenient point at which to address the topic of magnification. In this case we started with a specification which stated that the nominal magnification of the microscope should be 10 ×. We selected lenses that would yield this value, from a thin-lens perspective. Let's now examine the final design and determine an absolute value for the actual magnification. In order to be absolute, a definition of magnification must first be established. In this case it will be recalled that the microscope is being used to examine a 12-mm-diameter object. Without a microscope, that object would be viewed at the near point of the eye (10 in or 254 mm), where it would subtend an angle of 2.25° to the eye. If, using the computer model, we refocus the original microscope system such that its final image is formed 254 mm from the eye (to the left of the exit pupil), the diameter of that image is found to be 118.6 mm. Since the object diameter is 12.0 mm, the magnification in this case may be calculated as 118.6/12.0 = 9.88 × (see Fig. 5-13). While not the exact value set forth in the specification, in most cases this value would be close enough to the goal of 10 × to be considered acceptable. It is interesting to note that, when the microscope focus is set such that the final image appears at infinity, then that image is found to subtend an angle of 23° to the eye. The system angular magnification in this case is 23°/2.25° = 10.20 ×. Both magnification values are within 2 percent of the original specification. If the 10 × value were critical, it would first have to be more precisely defined. It could

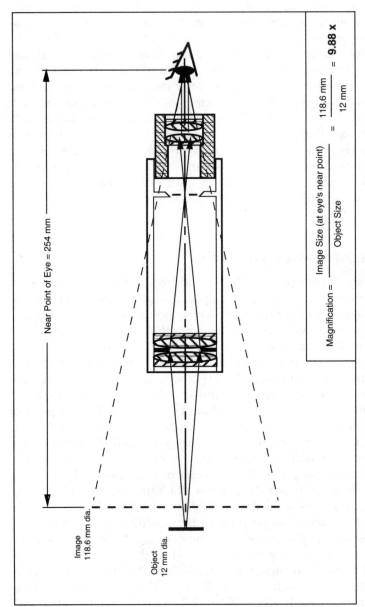

Near Point of Eye = 254 mm

Image
118.6 mm dia.

Object
12 mm dia.

$$\text{Magnification} = \frac{\text{Image Size (at eye's near point)}}{\text{Object Size}} = \frac{118.6 \text{ mm}}{12 \text{ mm}} = \textbf{9.88 ×}$$

Figure 5-13. The actual magnification of the 10 × (nominal) microscope can be determined by adjusting the computer model for an image at the eye's near point (254 mm) and tracing exact rays through the optical system. The size of this image, compared with the size of the object, gives the actual magnification of the system.

then be achieved by a change in the object distance, to produce a slight change in the magnification at the first image plane.

This magnification analysis demonstrates one of the major advantages of working with today's optical design computer systems, that is, the ability to generate actual results through the ray tracing process, as opposed to the use of generic formulas, which are often subject to interpretation and may or may not apply to the specific problem at hand.

5.10 Design of a Noncatalog System

While the catalog lens data base is certainly one of the more useful features of OSLO MG, there are many design tasks that can be accomplished outside that realm. This example will demonstrate one of these. Quite often, when reviewing optics related literature, we will find material that inspires us to generate a system design that includes this newly acquired information. For example, suppose data appear describing a new form of doublet lens offering many advantages over traditional forms. This might inspire us to start dreaming about the ultimate refracting telescope, suitable for our amateur astronomy activities. With a modern optical design package, it will be possible to generate and evaluate just such a design.

The literature will typically contain an illustration and a table of lens data such as that shown in Fig. 5-14. The major advantage claimed for this design is diffraction-limited performance over the visual spectral bandwidth. This is rather remarkable, since a conventional achromatic doublet of these dimensions will be limited by secondary color to the extent that its performance is nearly three times the diffraction limit. As is almost always the case, this performance level is not reached without paying a price. In this case that price would be the actual dollar value, the thermal sensitivity and the difficulty of manufacture, associated with the optical glass used for the second element (Schott FK54). Assuming we are willing and able to pay that price, let's proceed with the design effort.

A preliminary lens layout will allow us to establish the basic characteristics of our telescope. The objective lens will take the new form, it will have an aperture of 100 mm, a focal length of 1000 mm, and a half field of view of 0.8°. The overall telescope magnification will be set at 25 × for this example. This can be modified in use by simply changing eyepieces.

For a 25 × overall magnification we will need an eyepiece with a focal length of $1000 \div 25 = 40$ mm. The eyepiece half field of view will

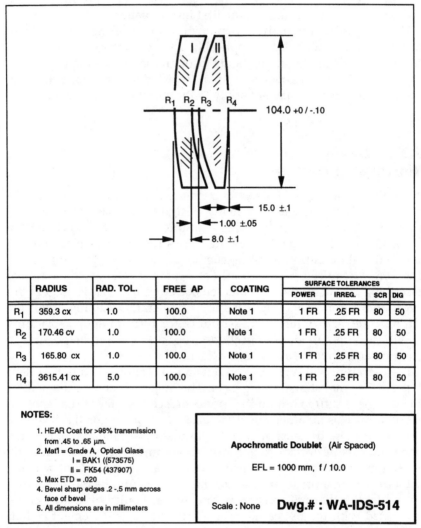

	RADIUS	RAD. TOL.	FREE AP	COATING	SURFACE TOLERANCES			
					POWER	IRREG.	SCR	DIG
R_1	359.3 cx	1.0	100.0	Note 1	1 FR	.25 FR	80	50
R_2	170.46 cv	1.0	100.0	Note 1	1 FR	.25 FR	80	50
R_3	165.80 cx	1.0	100.0	Note 1	1 FR	.25 FR	80	50
R_4	3615.41 cx	5.0	100.0	Note 1	1 FR	.25 FR	80	50

NOTES:

1. HEAR Coat for >98% transmission from .45 to .65 µm.
2. Mat'l = Grade A, Optical Glass
   I = BAK1 ((573575)
   II = FK54 (437907)
3. Max ETD = .020
4. Bevel sharp edges .2 -.5 mm across face of bevel
5. All dimensions are in millimeters

Apochromatic Doublet (Air Spaced)

EFL = 1000 mm, f / 10.0

Scale : None   **Dwg.# : WA-IDS-514**

Figure 5-14. Typical presentation of data describing a new, apochromatic doublet lens design.

be $0.8 \times 25 = 20°$. Designing an eyepiece is perhaps the ultimate example of reinventing the wheel. Optics literature abounds with established eyepiece designs. In this case, an orthoscopic eyepiece design, such as that found in *MIL Handbook 141*, will be consistent with the overall telescope quality that we desire. The resulting preliminary system layout is shown in Fig. 5-15. OSLO MG will be used to build a computer model and evaluate the performance of this telescope.

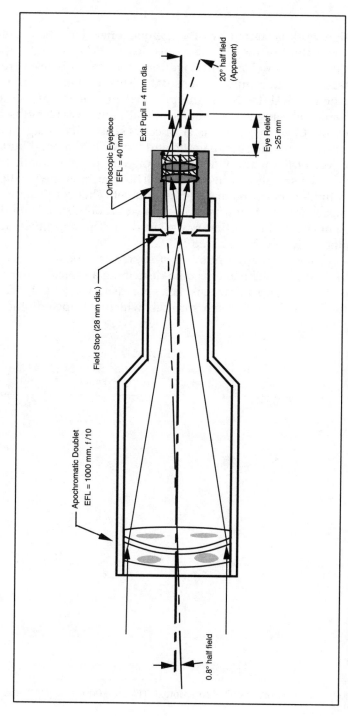

Figure 5-15. Starting layout of a 25 × telescope made from a 1000-mm apochromatic doublet, combined with a 40-mm orthoscopic eyepiece.

The design work is started by first opening a new lens data spreadsheet for evaluation of the objective lens design. After informing the computer that our new lens will have four surfaces, a blank spreadsheet appears, awaiting input of lens data. This is done by inserting the published lens data. Next, the paraxial properties of the lens (diameter and field angle) are entered. Finally, the wavelengths and weights are modified to reflect visual spectral sensitivity. Image quality is confirmed using the spot diagram and the MTF command buttons on the tool bar of the main window. The on-axis, polychromatic MTF curve for this lens is shown in Fig. 5-16. This confirms the predicted diffraction-limited performance of this doublet. This confirmation is critical, since it will be found in many cases that published data does not represent the real or optimum design data. The objective lens is stored temporarily under the name *TELE*.

For the eyepiece design, a number of sources are possible. Section 14.6 of *MIL Handbook 141* contains data describing a typical orthoscopic eyepiece. Opening up a new lens data spreadsheet in OSLO MG, the lens is specified as having seven surfaces, which will include the exit

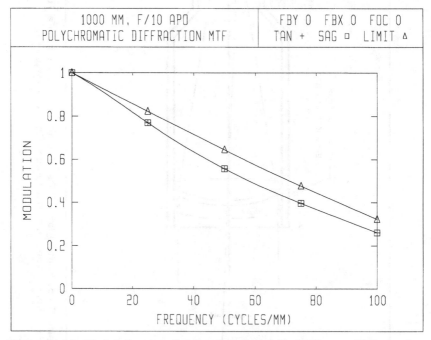

Figure 5-16. Modulation transfer function (MTF) for 1000-mm, *f*/10 apochromatic doublet.

pupil of the eyepiece. The data from the handbook are then entered into the lens data spreadsheet. The paraxial data then is inserted, including an *f* number of *f*/10, and a half-field angle of 20°. This lens is found to have a focal length of 2.54, as is indicated in the source data. It is a simple matter to invoke the *scale EFL* command to change the EFL to 40 mm. This eyepiece design is now evaluated by examining its aberration curves and is found to be in general agreement with the handbook data. From the lens data spreadsheet, all eyepiece lens surfaces are selected and then copied to the computer clipboard using the copy command.

The objective lens is then restored from the computer lens library and the eyepiece is added to it by copying it from the clipboard into the lens data spreadsheet. It will be noticed immediately that the eyepiece orientation is reversed (the image is to the right). This is easily remedied by selecting and reversing the eyepiece elements using the lens edit command. As was the case with the microscope in the previous example, we now have a system that has an output in the form of collimated light bundles. This must be accommodated by changing the general operating condition of the lens to the afocal mode. The quickest, and most telling method of evaluating image quality in this case is the MTF calculation. Clicking on the MTF toolbar button produces the polychromatic diffraction MTF curve shown in Figure 5-17. This confirms the diffraction-limited performance of our new telescope design.

While the design task at this point is essentially complete, it is possible to perform further analysis of this system using the wavefront analysis feature of OSLO MG. Activating the wavefront button on the toolbar generates the graphic shown in Fig. 5-18. This illustration indicates the condition of the wavefront as it emerges from the telescope eyepiece and passes through the system exit pupil. The output contains both a digital contour map, and a three-dimensional (3D) picture of the wavefront for three field angles. From this figure we conclude that the on-axis performance is essentially perfect, while some astigmatism is present at the 0.56° and 0.8° field points. For the maximum field angle we see about four waves of astigmatism, due primarily to the residual aberrations of the eyepiece. The stark reality of the changes that have taken place in the field of lens design in recent decades becomes dramatically apparent when we think about the lens designer of the 1950s, who hesitated to trace a single additional ray through the system in question because of the amount of work and time involved. Today, with a single click of the computer's mouse button, we can instantly view a complete wavefront analysis. We have indeed come a long way.

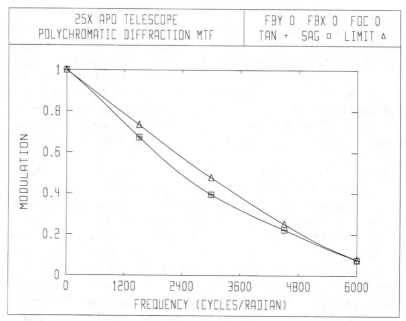

Figure 5-17. Modulation transfer function (MTF) for a 25 × apochromatic telescope.

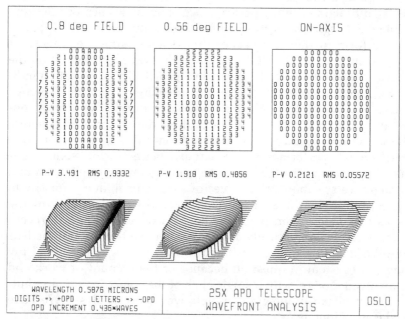

Figure 5-18. Wavefront analysis for a 25 × apochromatic telescope as produced by OSLO MG.

5.11   Review and Summary

This chapter has dealt primarily with the application of a modern, low-cost software package to the design and analysis of optical systems. In order to create a sense for the significance of these modern optical design packages, an historic perspective was presented, dealing with the field of lens design in general and the development of the OSLO program by Sinclair Optics in particular. The key point that has surfaced from this historic review has been the shifting of responsibilities that has taken place within the field of optical engineering in recent years. This has resulted in a considerable overlap in the responsibilities of the optical engineer with those of the lens designer. While a familiar title, one that has existed for many years, this chapter has attempted to define the work of the optical designer in terms that are consistent with today's workplace conditions.

Several lens design examples have been presented with the aim of illustrating the broad scope of work that can be handled, the flexibility of these programs, and the ease with which they can be used. Perhaps the most significant fact is that these programs can and will be used by a broad cross section of those involved in the science of optics. What was once the exclusive domain of the lens designer has become an area that can and should be explored by many. At the same time, this new approach to optical design allows lens designers to expand their work into areas that were previously not accessible to them. The net result of all this will be increased productivity and better solutions to a wide range of optical design problems. It also makes the work of all involved in the field of optical engineering more challenging and more exciting.

Note: The OSLO MG optical design software package is available from

Melles Griot
 1770 Kettering Street
 Irvine, CA 92714
 Phone: (800) 835-2626
 Fax: (714) 261-7589

Other versions of the OSLO programs are available from

Sinclair Optics, Inc.
 6780 Palmyra Road
 Fairport, NY 14450
 Phone: (716) 425-4380
 Fax: (716) 425-4382

6
Primary Lens Aberrations

6.1 Introduction

It is the function of a lens to collect light from a point on the object and to focus that light at a corresponding point (a conjugate point) on the image. In nearly all cases, the lens will fail at this task, in that there will be some residual error in the precision with which it collects, refracts, and focuses this light. Rather than a true point image, the lens will produce a blur circle, or a spot. It is the function of the optical designer to ensure that this spot size is sufficiently small to allow the lens to produce the required resolution, or image quality. These errors in the lens's ability to form a perfect image are referred to as *lens aberrations*. There are seven primary aberrations that should be taken into consideration when a lens system is being designed or evaluated. By understanding the basic characteristics of these seven primary lens aberrations, the optical engineer will be better equipped to specify and evaluate the potential image quality of an optical system. The following paragraphs will describe the seven primary lens aberrations, and discuss some of the salient points of each.

6.2 Spot Diagram and Radial Energy Distribution

There are a number of methods available for the visualization and evaluation of the image quality that will be produced by a lens or optical system. For purposes of this discussion of primary lens aberrations, we will be dealing with two of the more commonly used methods; the spot diagram and the radial energy distribution calculation. A spot diagram is generated by tracing a relatively large number of rays (usually 200 to 300) from a common object point, through the lens system, and plotting the spot that results due to the intersection of those rays with the image plane. This spot diagram represents a reasonably accurate simulation of the image of a point source object, such as a star, that will be produced by the lens. If the lens were perfect, all rays would intersect at the same point and the spot diagram would be just that...a point. (This purely geometric analysis neglects the effects of diffraction, which will be discussed later in this chapter.) When residual aberrations are present, the spot takes on a finite size and shape. Analysis of that spot makes it possible to determine a great deal about the ultimate image quality of the lens.

The most common and useful form of spot diagram analysis involves starting at the center of the spot, gradually increasing the spot radius, and counting the number of rays that are contained within that radius. The resulting curve, showing the percent of energy encircled vs. the spot radius, is referred to as the *radial energy distribution* (RED) curve. Figure 6-1 shows a typical spot diagram, generated by tracing 240 rays through a well-corrected lens. The corresponding RED curve for that spot is also shown. From this spot diagram data it can be concluded that the image formed by this lens, of any point on the object, will be a spot with a radius of about 0.024 mm. Within that image, it will not be possible to resolve detail that is finer than that spot size. This form of image analysis (the spot diagram and RED) will be used throughout the remainder of this section to describe the impact that the various aberrations will have on image quality.

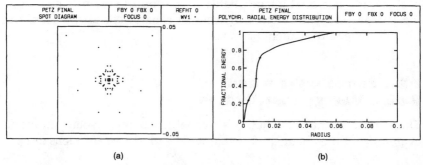

(a)   (b)

Figure 6-1. On-axis spot diagram (*a*), and RED curve (*b*) for a well-corrected Petzval lens.

6.3 Spherical Aberration

There are three basic aberration types: on-axis, off-axis, and chromatic. The principal on-axis lens aberration is *spherical* aberration. This is the imaging error found when a lens is called on to focus an axial bundle of monochromatic (single wavelength or color) light. In the presence of spherical aberration, each zone, or annulus, of the lens aperture will be found to have a slightly different image distance. The result of this can be seen in Fig. 6-2d, which illustrates the presence of spherical aberration in a simple, planoconvex lens. In this case, spot diagram analysis has been done at three separate focus positions; position 1 is the paraxial focus. At the paraxial focus, all rays passing through a zone near the center of the lens will be accurately focused. However, as rays from zones farther from the lens center are considered, they will be focused progressively short of the paraxial focus. The farther the rays are from the center, the greater will be this error in focus. This lack of a common focus, or image distance, for all zones of the lens aperture is *spherical aberration.*

We see in Fig. 6-2d that there is a point short of the paraxial focus where the blur circle, or spot size due to spherical aberration, is minimized. In the diagram that point has been designated 3. Point 2, which is midway between points 1 and 3, has also been selected for analysis.

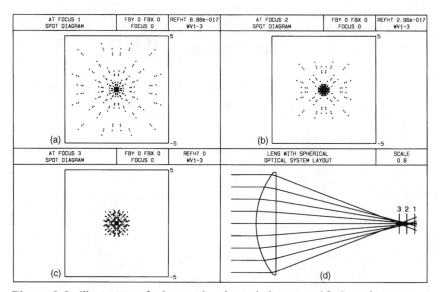

Figure 6-2. Illustration of a lens with spherical aberration (d). Spot diagrams are shown for three focus positions: paraxial (1), minimum spot size (3), and the midpoint between those two (2).

Figure 6-2*a*, *b*, and *c* are the spot diagrams that correspond to focus positions 1, 2, and 3, respectively. A good sense of the image quality of this lens in the vicinity of the paraxial focus can be gained by a careful study of these spot diagrams.

Figure 6-3 shows the RED curves corresponding to the three spot diagrams that were shown in Fig. 6-2. Analysis of these RED curves will yield a pretty good indication of relative image quality at each point of focus. In general, the smaller the spot, the better the image quality. Reviewing the three RED curves, it appears that the focus position for minimum spot size (3) would represent the point of best image quality. However, it can also be seen that, for the 25 percent encircled energy level, the spot radius at the middle focus position (2) is about half the size of the corresponding spot at the minimum spot size focus (3), while at the 70 percent level, both of these curves have essentially the same value. In the end analysis, there would probably be very little difference in image quality between these two focus positions. One reasonable way to choose between two RED curves is to select that curve with the least area to its left as representative of optimum image quality. In almost all applications where spherical aberration is present, the overall image quality will be best when the lens is focused away from the paraxial focus, close to the point of minimum spot size. In addition to determining the point of best focus, this analy-

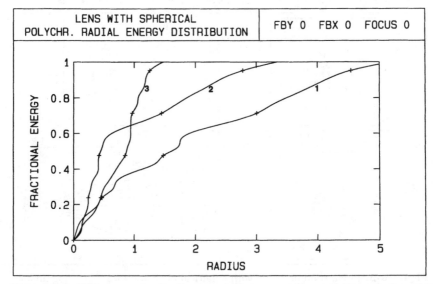

Figure 6-3. RED curves for the three spot diagrams at focus positions 1, 2, and 3 shown in Fig. 6-2.

sis also gives a good indication of the sensitivity of focus for this particular optical system.

In the case of a simple lens element, the number of methods available to control spherical aberration are limited. First, while maintaining the power of the lens constant, its shape may be altered such that the amount of spherical aberration is minimized. This is referred to as *bending the lens*. A second approach to the reduction of spherical aberration involves the increase in the index of refraction n of the lens material. In going from a standard glass ($n = 1.5$), to a very high-index glass ($n = 1.9$), the amount of residual spherical aberration can be reduced by about one half. While the spherical aberration can be minimized by lens bending and index change, it cannot be eliminated completely when a real object and image are involved. For this reason it is important for the optical engineer to establish just how much residual spherical aberration is allowable, while still producing an acceptable system design. In order to make this determination the optical engineer must develop some sense of just how the optical system will function, and what levels of image quality will be compatible with other system components (such as the detector) and with the overall performance requirements of the system.

6.4 Coma

Coma is an aberration that affects off-axis light bundles in a manner quite similar to the way in which spherical aberration affects axial bundles. As is shown in Fig. 6-4, when an off-axis bundle is incident on a lens that is afflicted with coma, each annulus, or zone of the lens aperture, focuses onto the image plane at a slightly different height and with a different spot size. The result is an overall spot that is basically triangular in shape, having a bright central core at its base, with the triangle-shaped flare extending toward the optical axis of the lens. Figure 6-5 shows a spot diagram for an image point where coma is the predominant aberration. Also shown is the RED curve for this spot. Because this spot is not symmetrical, the centroid for the RED calculation must be selected statistically.

For a pair of simple lenses arranged symmetrically about the system aperture stop, as in a relay lens system, or for a complex lens that has a reasonable degree of symmetry about its aperture stop, there will be a significant reduction in the amount of coma found. This important characteristic is used in the design of many lenses and optical instruments, such as the double-gauss lens, borescopes, and submarine periscopes. The residual coma in a lens system will usually be mixed

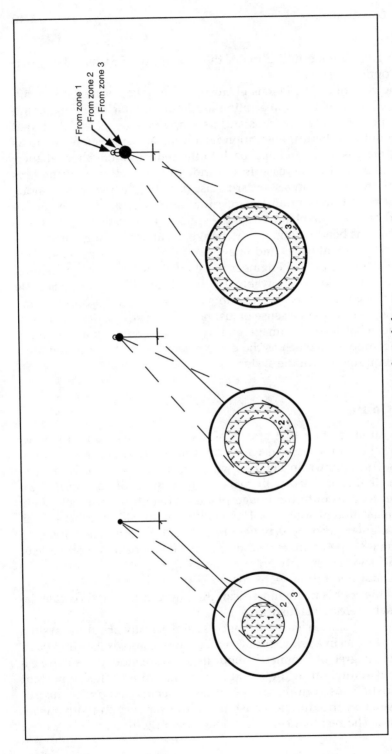

Figure 6-4. *Coma* is an off-axis aberration similar in many ways to spherical aberration. The formation of a blur due to coma is illustrated here, as light from three zones of the lens aperture forms spots of three sizes, at varying heights on the image plane.

From zone 1
From zone 2
From zone 3

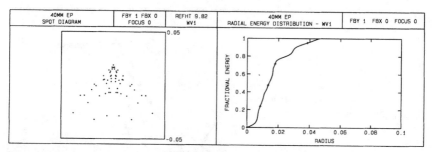

Figure 6-5. Spot diagram and RED curve for an image with coma.

with other off-axis aberrations, making its individual contribution to final image quality difficult to evaluate.

6.5   Field Curvature

In most optical systems it is required that the final image be formed on a flat surface. Unfortunately, it is the natural tendency of most optical systems to form that image on a curved surface. The nominal curvature (1/radius) of that preferred image surface is referred to as the *Petzval curvature,* or *field curvature* of the lens. For a simple lens this curvature will be equal to approximately two-thirds of the lens power, or the radius of the Petzval surface will be approximately 1.5 × the focal length of the lens. When the lens is free of all other aberrations, an essentially perfect image will be formed on the curved Petzval surface. When astigmatism is present (which is most often the case), the Petzval surface has little real significance as far as the actual image quality of the lens system is concerned.

6.6   Astigmatism

When astigmatism is present in a lens system, off-axis fans of rays that are orientated tangentially and sagittally at the lens aperture will tend to focus on two different curved surfaces. Figure 6-6 shows two such fans of rays, one in the tangential (Y-Z) plane, the other in the sagittal (X-Z) plane, passing through a simple lens, and illustrates how these ray fans will be focused by a lens when astigmatism is present. It can be seen here, and also from the corresponding spot diagram in Fig.

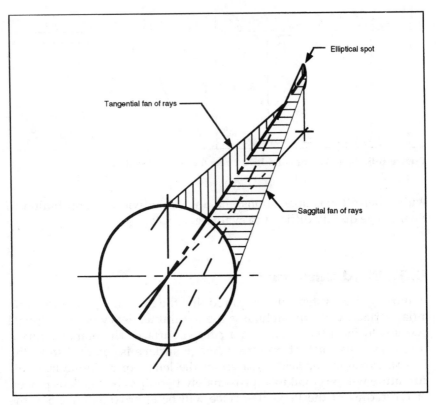

Figure 6-6. *Astigmatism* is that off-axis aberration caused by the lens having a different focus for the tangential fan of rays vs. the sagittal fan. The result will most often be an elliptical rather than circular image of a point source.

6-7, that the presence of astigmatism will cause the ideal circular spot to be blurred into an elliptical, or astigmatic, shape. Since this elliptical spot has significant differences between its horizontal and vertical dimensions, it follows that at any given focus, the resolution will vary, depending on the orientation of the lines in the object being imaged. For this same basic reason, the RED calculation is not meaningful and is seldom performed on an astigmatic spot.

The field curves shown in Fig. 6-7 represent the most common method of illustrating and evaluating the aberrations of field curvature and astigmatism. These curves represent a cross section through half the image surface, from the optical axis out to the edge of the field. In the case illustrated here, the system has been focused on the Petzval surface at the maximum field angle. The tangential and sagit-

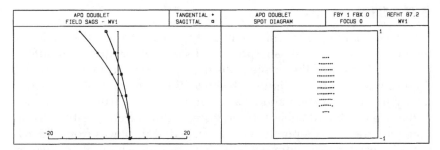

Figure 6-7. Field sag curves and spot diagram for an image with astigmatism.

tal field curves are plotted relative to that reference. If we consider the case where the image being formed is of a spoked wheel, with its hub centered on the optical axis, the rim of the wheel will be in focus on the tangential image surface, while the spokes will be in focus along the sagittal surface. Astigmatism is, by definition, the difference between the tangential and sagittal field curves. If the tangential and sagittal surfaces are coincident, then the lens is said to be free of astigmatism and the curved image will be formed on the Petzval surface. When astigmatism is present, the distance from the Petzval surface to the tangential field curve will be three times the distance to the sagittal field curve (see Fig. 6-7). In most cases it is not possible to correct field curvature and astigmatism to zero, but satisfactory image quality is most often achieved by balancing residual astigmatism with inherent field curvature.

Another useful approach to the correction of field curvature and astigmatism is to first correct the lens such that there is no astigmatism and the image is formed on a surface whose curvature is equal to the Petzval curvature of the lens. Once this has been accomplished, a field flattener lens can then be added in close proximity to the final image. The Petzval curvature of the field flattener is made equal to and opposite in sign to the Petzval curvature of the basic lens. The net result will be a lens that is free of astigmatism and has zero field curvature. The drawback to this approach (remember, there is always a tradeoff) lies in the fact that the field flattener must be located very close to the image plane. In the worst case, some mechanical interference condition might exist, making this approach physically impossible. In another case, the proximity of the field flattener to the image plane means that its lens surfaces must be meticulously clean, free of scratches, dirt, and dust, which, if present, would appear superimposed on the image.

6.7 Distortion

Distortion is a unique aberration, in that it does not affect the quality of the image in terms of its sharpness or focus. Rather, distortion affects the overall shape of the image, causing it to depart from a true scaled duplicate of the object. Figure 6-8 shows three lenses; the first (left) is totally free of distortion and as a result, produces a true reproduction of the "checkerboard" object. If the system suffers from positive distortion, then the off-axis points will be imaged at distances from the axis that are greater than nominal, creating the pincushion effect seen in Fig. 6-8 (center). On the other hand, if the system exhibits negative distortion, the resulting image will assume a barrel shape as is seen in Fig. 6-8 (right). For most visual systems, distortion errors in the 5 to 10 percent region are usually considered acceptable. For camera and projection lenses, distortion values in the 1 to 2 percent range are common.

Note: The five aberrations presented to this point have been monochromatic aberrations, generally computed at the central wavelength of the lens system. If the lens is to be used over an extended spectral bandwidth, the following chromatic aberrations must also be considered.

6.8 Axial Color

In an earlier section dealing with dispersion it was stated that the index of refraction for all optical glasses will vary as a function of wavelength. That index will be greater for shorter wavelengths (blue), and the rate at which the index changes will also be greater at those shorter wavelengths. In a simple lens this results in each wavelength being focused at a different point along the optical axis. While this chromatic spreading of the light by a prism is referred to as *dispersion,* when it occurs in a lens as described here, it is referred to as *primary axial color.* Figure 6-9a illustrates a simple lens focusing an axial bundle of white light. If the focus is set for the middle of the band as shown, the blur circle will consist of a green central core with a halo of purple (red plus blue) light surrounding it. Except in very unusual cases, such as laser systems or other nearly monochromatic systems, axial color is an aberration that must be dealt with in order to achieve usable image quality. This can be most easily accomplished by converting the simple lens into an achromatic doublet as is shown in Fig. 6-9b. Here, two glass types and lens powers are selected such that the primary axial color has been corrected by bringing the two extreme wavelengths (red and blue)

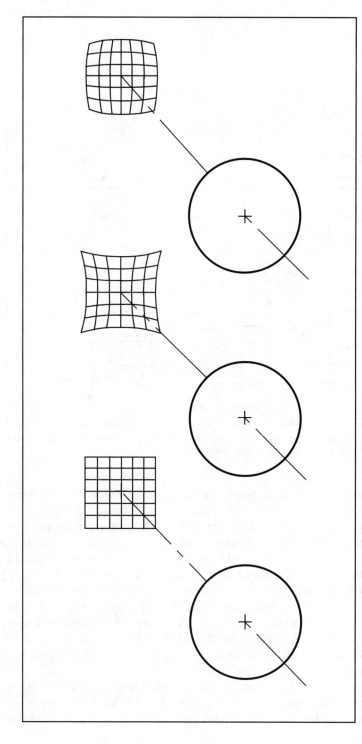

Figure 6-8. Distortion is present when the focal length of the lens varies as a function of field angle or image size. In this illustration, the lens on the left is free of distortion, the lens in the center has 16 percent positive (pincushion) distortion, the lens on the right has 10 percent negative (barrel) distortion.

139

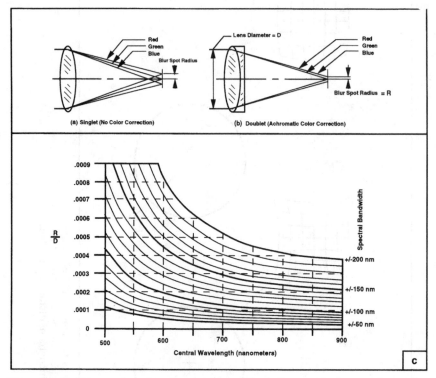

Figure 6-9. When a lens has been achromatized, the residual secondary color can be estimated using this chart (bottom panel).

to a common focus. In most cases, the substantial reduction in blur circle size due to achromatization will result in a satisfactory design.

When this state of achromatization is achieved, the primary color is said to be corrected. The remaining chromatic aberration is referred to as *secondary color*. Secondary color is often found to be the limiting axial aberration in a lens system that has been otherwise carefully designed and optimized. Because of this, it is quite useful to be able to estimate the amount of secondary color that might be expected after all the primary aberrations have been corrected. Figure 6-9c is a plot that makes it possible to estimate the residual blur circle size due to secondary color, assuming we know the diameter of the lens and the spectral bandwidth that is being imaged.

When an optical system contains a series of lenses, all of which have been corrected for primary color, it is valid to add the diameters of all lenses (based on the size of the axial bundle) and use the sum of the diameters as D when using Fig. 6-9c. The standard submarine periscope is an extreme example of this condition. In that case we might typically have an objective lens that is 42 mm in diameter, followed by six relay lenses, each with a diameter of 100 mm. All these lenses are achromatized, corrected for primary color over the visual spectrum. If we assume the spectral bandwidth to be 550 ± 120 nm, we can apply these values to Fig. 6-9c and will find that the corresponding value for R/D is equal to about 0.00034. Knowing the cumulative lens diameter D to be 642 mm, we can then solve for the blur spot radius R, which we find to be 0.22 mm. Quite surprisingly, this blur circle of nearly 0.5 mm diameter due to secondary color does not detract greatly from the visual performance of the submarine periscope. However, when high-quality photography is the goal, it does become a problem, making it necessary to introduce a filter to reduce the spectral bandwidth being imaged.

When deemed absolutely essential, it is possible to design the basic lens system such that secondary color is significantly reduced or eliminated. Such a design will have a common focal length for three wavelengths rather than just two, and is referred to as an *apochromatic design*. Consistent with the "you don't get something for nothing" philosophy, this step will invariably lead to a very expensive lens system, one that will often be found to have poor environmental characteristics.

6.9 Lateral Color

The second chromatic aberration (and the last of the seven primary lens aberrations) is lateral color. For on-axis light bundles, the central ray in that bundle will coincide with the optical axis of the lens. For off-axis light bundles the corresponding central ray is called the *chief*, or *principal*, ray. The height of the chief ray at the image plane defines image size. When lateral color is present in the lens system, the refraction of the the chief ray will vary as a function of wavelength, causing each wavelength to be imaged at a slightly different height on the image plane. The result is a chromatic, radial blur for off-axis image points. In the case of a simple lens, with the chief ray passing through its center, there is little refraction of that ray and therefore little lateral color. Any system that is symmetrical about the point where the chief ray crosses the optical axis (the aperture stop) will have little or no lateral color because of the tendency of that aberration to cancel itself as the chief ray traverses the symmetrical halves of the system.

On the other hand, the eyepiece is a classic example of a lens form that requires large amounts of chief ray refraction that is not symmetrical about the aperture stop. As a result, in most eyepiece designs, lateral color is found to be a major contributor to degradation of off-axis image quality. Figure 6-10 illustrates the chief ray path, first through a simple lens and then through an eyepiece. The presence, or lack of chief ray refraction and the resulting lateral color is shown in each case.

6.10 Aberration Curves

Exposure to various textbooks and reference material dealing with optical system and lens design will invariably involve the presentation of aberration curves in one form or another. Aberration curves represent a method of plotting and presenting lens performance data that can be very useful. In this section we will review the set of typical aberration curves shown in Fig. 6-11, highlighting some of the more interesting and valuable information that can be derived from them. In this example, aberration curve data have been generated by tracing three tangential and three sagittal fans of rays through the lens. Each tangential fan contains nine rays, equally spaced from the top to the bottom of the aperture, while each sagittal fan contains four rays, equally spaced from the edge of the aperture to the center (for the sagittal fan, left to right symmetry makes it unnecessary to trace both sides of the aperture). Each ray in a tangential fan has a Y value at the aperture stop with a relative value between $+1.0$ and -1.0. Each ray has a corresponding Y value as it intersects the image plane. The aberration curves are generated by plotting these values relative to each other.

In the case of the sagittal fans, it is the X value of the ray at the aperture stop vs. the X value of the ray at the image plane that is plotted. Typically, tangential and sagittal ray fans are plotted for the on axis case, for 0.7 of the full field and for the full-field position. This is a total of 38 rays that must be traced through the system in order to generate a full set of aberration curves. Tracing of these 38 rays must be repeated for as many wavelengths as are of interest to the designer, usually three. Prior to the availability of the modern computer the lens designer would go to great lengths to reduce the number of rays that had to be traced. Useful results could often be obtained by tracing as few as seven rays. Today, that problem does not exist. For example, with the OSLO software package on a basic PC, it requires just a single keystroke and a matter of seconds for the designer to generate a complete set of aberration curves such as those shown in Fig. 6-11. Here we see, not only the basic aberration curves on the left, but also separate curves

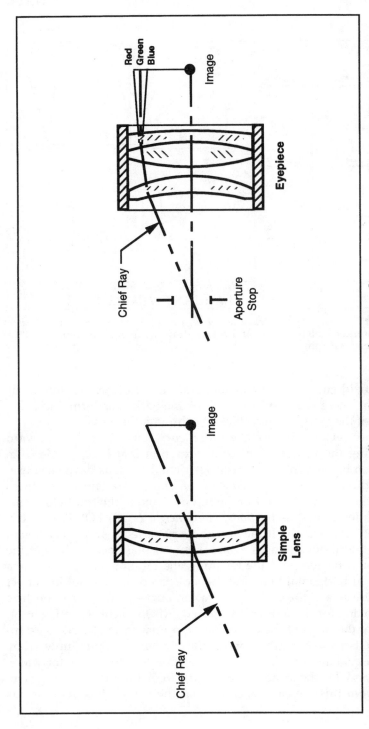

Figure 6-10. Lateral color is the dominant off-axis chromatic aberration. It results when the chief ray refraction is variable, as a function of wavelength. When little chief ray refraction occurs (left), there is little lateral color. When considerable chief ray refraction occurs (right), then there will be considerable lateral color.

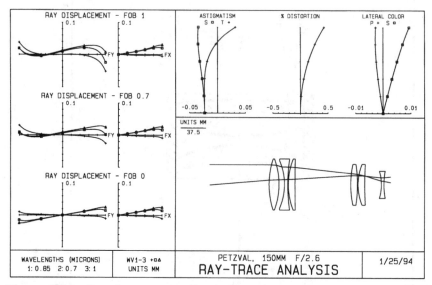

Figure 6-11. Complete ray trace analysis, including aberration curves, field curves, distortion, lateral color, and a lens drawing, as output from the OSLO Series 2 lens design program.

showing field curvature, distortion, and lateral color. Finally, in the lower right, there is a lens layout of the the basic lens form, including the paths of the marginal and chief rays through the lens.

A cursory review of these aberration curves will be helpful toward appreciating the vast amount of information that is available from them. It can be seen that, for this example, all ray fans have been executed in three wavelengths. Aberration curves are shown for three field positions, on axis, at 0.7 of the full field and at the full-field angle. These are designated as *fractional object height* (FOB) FOB 0, FOB 0.7, and FOB 1, respectively. If the system were perfect, completely free of residual aberrations, then all Y and X ray displacement values at the image would be zero and the resulting aberration curves would be a series of flat horizontal lines. Whenever a fan of rays is not in perfect focus, there is a slope of the aberration curve at the zero aperture point. Looking for the moment at the axial (FOB 0) curves in Fig. 6-11, we see that the slope of the curve for the primary wavelength is essentially zero. For wavelengths 2 and 3, the curves are sloping upward by about an equal amount. This is an indication that the focus for wavelengths 2 and 3 is the same, but not equal to that of the primary wavelength. From this we can conclude that the primary axial color has been corrected. From these curves it can be estimated that the maximum aberration (ray displacement) at wavelengths 2 and 3 is about 0.02 mm, which would be the approximate radius of the on-axis blur circle. The fact that the aberration curve for the primary wavelength has a slight curve downward at its end indicates the presence of some spherical aberration, but it can be concluded that on-axis image quality will clearly be dominated by the residual secondary color.

Looking at the off-axis curves, we can see that the sagittal curves are all essentially the same, regardless of field angle. We conclude that the sagittal field is flat, which is confirmed by the astigmatism curves seen elsewhere in Fig. 6-11. In the case of the off-axis tangential curves, it can be seen that as the field angle increases, the state of color correction is disturbed. It is interesting to note that, for the off-axis points, while the tangential ray displacements are greater over half of the aperture, they are actually less for the other half. We can estimate from this that the spot size at the maximum field angle will be about the same size as the on-axis spot. The result then should be reasonably consistent image quality as a function of field angle.

Looking at the curves for astigmatism, distortion, and lateral color (upper right, Fig. 6-11), it should be noted that the scale for plotting these curves has been selected by the computer to fill the allotted space. The actual amount of astigmatism and lateral color, and the impact of these on image quality, is best judged by examining the aberration curves or by executing spot diagrams on the system. The distortion curve yields a direct indication of the percent distortion that is present for any field angle. Finally, the accompanying lens picture (lower right, Fig. 6-11) will often be useful in helping to identify the source of aberrations and in suggesting possible approaches to improving the design.

6.11 Spot Diagram Analysis

Returning for the moment to the subject of spot diagrams, another very convenient feature of the OSLO program, that ties into the aberration curves just discussed, is the spot diagram analysis. Again, with just a single keystroke, the system will generate the information shown in Fig. 6-12. As with the aberration curves, the spot diagram analysis is performed on-axis, at 0.7 of the full field and at the full-field angle. On the left in Fig. 6-12 we can see the three spot diagrams that correspond to the three sets of aberration curves shown in Fig. 6-11.

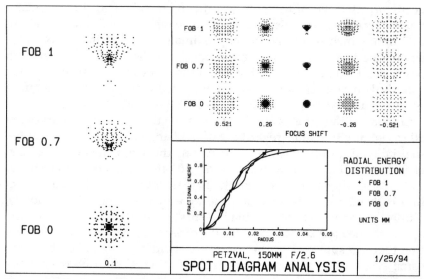

Figure 6-12. Complete spot diagram analysis, including through focus spot diagrams and RED curves, as output from the OSLO Series 2 lens design program.

In the upper right (Fig. 6-12), these three spots are shown again (at reduced scale), along with additional spot diagrams for the three field positions, at four different focus positions. This set of spot diagrams will give the designer a good sense for the system's sensitivity of focus. Finally, in the lower right (Fig. 6-12), we see the RED curves that correspond to the three spot diagrams at best focus. Noteworthy in this particular design is the fact that the three RED curves, for the three field positions, are nearly identical. This fact indicates a lens with very uniform and consistent image quality over its entire field of view. It will be recalled that, on the basis of aberration curves, we had estimated the on-axis blur circle radius would be about 0.02 mm and that spot size across the field would be quite uniform. From the RED curves we find that approximately 80 percent of the energy is contained within that 0.02 mm radius for all field positions, confirming that preliminary estimate.

There is one factor that can invalidate all conclusions drawn from analysis based on aberration curves and spot diagrams...that would be diffraction. Since all aberration and spot diagram information is generated by geometric ray tracing, it obviously does not take into account diffraction effects. It will be recalled from the discussion in Chap. 3

that, regardless of how well corrected an optical system might be, a blur circle of finite size (the Airy disk) will be produced as a result of diffraction effects at the aperture. That blur circle size is dictated by the f number of the optics and the wavelength of the energy being imaged according to the following formula:

$$\text{Blur circle radius} = 1.22 \times \lambda \times f \text{ number}$$

In the case of the lens used to generate the data shown in Fig. 6-12, the wavelength (λ) is 0.00085 mm and the f number is 2.6. The resulting blur circle radius due to diffraction would be

$$1.22 \times 0.00085 \times 2.6 = 0.0027 \text{ mm}$$

Because the airy disk radius is about $\frac{1}{10}$th the spot size that is indicated by geometric analysis, it can be concluded that the diffraction effects in this case would be negligible. When the spot size is found to approach the size of the airy disk (5 $\times$ or less), it is possible to generate RED data with the OSLO program that combine the geometric and diffraction effects.

The seven primary lens aberrations are sensitive in varying degrees to changes in the lens aperture size and the image size or field angle. From Fig. 6-13 it can be seen how the spot size due to these aberrations will be affected as the aperture and field size of a lens are modified. For example, if the lens diameter were doubled, the spherical aberration would increase by a factor of $2^3 = 8 \times$; coma, by 4 $\times$; and field curvature and axial color, by 2 $\times$. Distortion and lateral color would not change.

6.12 Review and Summary

This completes our review of the seven primary lens aberrations, along with a brief look at spot diagrams, radial energy distribution, and aberration curves. It is the function of the optical designer to evaluate the impact that residual aberrations will have on final system performance and to adjust the configuration of the lens system such that satisfactory performance will result. This topic once again demonstrates quite dramatically the impact that the modern computer system and available optical design software has had on the scope of work and design responsibilities that are shared today by the optical engineer and the lens designer, i.e., the optical designer.

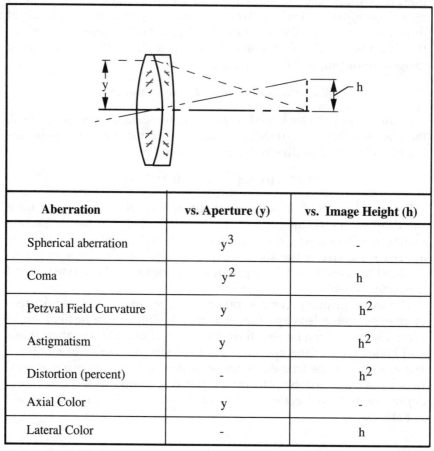

Aberration	vs. Aperture (y)	vs. Image Height (h)
Spherical aberration	y^3	-
Coma	y^2	h
Petzval Field Curvature	y	h^2
Astigmatism	y	h^2
Distortion (percent)	-	h^2
Axial Color	y	-
Lateral Color	-	h

Figure 6-13. The blur circle diameter due to the primary aberrations will vary in size as a function of aperture radius y and image height h.

7

Optical Components

7.1 Introduction

All optical instruments, or systems, will contain a series of optical components. Each of these components has a specific function involving the collection, refraction, reflection, or deviation of the light coming from an object. This chapter is intended to give the reader a basic level of familiarity with this family of optical components. Armed with this familiarity, the optical engineer will be better equipped to specify, design, and analyze a broad range of optical systems. The three components that are most frequently encountered are the lens, the mirror, and the prism. Each of these will be covered here in some detail, including a discussion of typical applications and performance limitations.

7.2 The Lens

The simple lens is the primary component of nearly all optical systems. When a bundle of light from a point on the object is incident on a positive lens, that bundle will be refracted such that it is brought to a focus and an image of the object point is formed. A simple, single lens element can be described completely, and it can be thoroughly analyzed, if the following detailed parameters are known:

- Lens material
- Radius of both surfaces
- Lens thickness
- Lens diameter

In those cases where a substantial spectral bandwidth is to be imaged, it will be found that, as a result of the dispersion of the glass, the image quality of the single element lens will not be adequate. In the majority of cases it will be required that the lens be achromatized. This will lead to a two element, or doublet form of the simple lens. The two elements of the doublet may be cemented, or they may be air-spaced. The arguments in favor of the cemented doublet over the air-spaced form are many and often times convincing. The cemented surfaces will require less stringent tolerances and will not require antireflection coatings. Also, the need for an intermediate lens spacer will be eliminated, as will the potential for wedge in the airspace between the two elements. On the other hand, the air-spaced doublet will generally deliver better image quality, while obviously eliminating the need for the cementing operation. Depending on the size of the elements, cementing can sometimes be a problem, producing mechanical strain and deformation of the elements that can seriously degrade image quality. While none of these factors is conclusive, all should be considered when deciding between the cemented and the air-spaced doublet.

One final lens type in the genre that can be reasonably referred to as *simple lenses* is the doublet with a split-crown element. In the achromatic doublet, the positive element is referred to as the *crown*, while the negative element is referred to as the *flint*. These designations are based on the optical glass types typically used for each. When it is found that the doublet produces poor image quality, due to residual spherical aberration, it may be possible to improve that performance substantially by splitting the crown into two elements, each having approximately one half of the total required positive lens power.

7.3 A Typical Lens Application

To demonstrate these simple lens types and their limitations, let's consider a typical application and generate some actual examples of how these simple positive lens types might perform. For example, assume we wish to design a basic, low-power (10 ×) telescope to be used for visual observation purposes. A variety of overall system considerations (including size and weight) might lead us to conclude that a 50-mm diameter lens with a focal length of 400 mm would serve well as an objective lens for such a telescope. Starting with a catalog lens, or a custom design, it will be quickly determined that a simple crown glass element, meeting these specifications, will produce a blur spot with a radius of 0.074 mm, due primarily to chromatic aberration over the visible portion of the spectrum.

This seems intuitively to be an unacceptably large spot size...let's see if we can confirm that. We had stated that the application involved here was a telescope for visual observation, with a magnification of 10 ×. It is a fact that the typical human eye is able to resolve detail with an angular subtense of about 1 arc minute (60 arc seconds). With the aid of a properly designed 10 × telescope, that resolution capability will be improved to $\frac{1}{10}$th that, or about 6 arc seconds. Now, the 0.074-mm blur of our simple 400-mm objective represents an angular blur (in object space) of about 37 arc seconds. Since our goal is resolution of about 6 arc seconds, or about $\frac{1}{6}$th of that, this confirms the fact that this level of image quality is not acceptable. In other words, using a 10 × telescope with this simple element for an objective lens, the viewers angular resolution would be about 37 arc seconds, only slightly better than with no telescope at all. While objects being viewed would appear 10 times larger, the image quality would be such that little additional detail in the object would be resolved.

Our next step in the design process would be to select (or design) an achromatic doublet with a diameter of 50 mm and a focal length of 400 mm. The resulting optimum cemented doublet design is found to have a blur spot radius of 0.003 mm, due mainly to residual secondary color, along with a small amount of residual spherical aberration. This 0.003-mm spot radius corresponds to an angular resolution of 1.5 arc seconds, which is well within our ultimate resolution goal of 6 arc seconds. In other words, with this cemented achromat as an objective lens, the telescope will deliver greater resolution than the eye is capable of. This may seem like a case of overkill, but it is actually a desirable condition, in that it allows for some degradation of lens performance due to manufacturing and assembly tolerances. In fact, it might be worthwhile to explore the possibility of further improvement of image quality by using an air-spaced doublet.

Changing the design to an air-spaced form provides one additional curvature to be used as a variable in correcting residual aberrations (for a doublet of this speed, the airspace itself is not found to be a useful variable). In the air-spaced design we find that the residual spherical aberration can be reduced to essentially zero, and the lens resolution is now limited strictly by the residual secondary color. The blur spot radius for the air-spaced doublet is reduced to about 0.0027 mm, which corresponds to an angular resolution of 1.3 arc seconds. In this case, legitimate arguments can be made for the advantages of both the cemented and the air-spaced form of achromatic doublet. In the end analysis, the cost benefits of the cemented form would probably prevail.

It was mentioned earlier that the next level of complexity for a lens of this type would involve splitting of the crown element into two components, each having reduced lens power. The advantage to be realized from this approach is primarily a reduction of the residual spherical aberration. It has been shown that the spherical aberration can be essentially eliminated by going to an air-spaced doublet, making the split crown approach not applicable in this case.

Figure 7-1 shows a scale drawing of the 50-mm-diameter, 400-mm-focal-length lens in the single element form. Also shown is the corresponding spot diagram, with a nominal (RMS) spot radius of 0.074 mm. This spot radius subtends an angle of 37 arc seconds in object space, which may be considered the limit of angular resolution for this lens. The spot size in this case is due primarily to the fact that this lens has not been color-corrected; it is not an achromat. Figure 7-2 shows a second lens, a cemented achromatic doublet with the same diameter and focal length as the single element. In this case the corresponding spot diagram has a radius of 0.003 mm, with an angular subtense of just 1.5 arc seconds. (Note the 20 × change in spot diagram scale between Figs. 7-1 and 7-2.) In this lens the choice of optical glass types, combined with the distribution of lens power, has resulted in an achromatic design. The air-spaced doublet shown in Fig. 7-3 has a slightly smaller spot with a radius of 0.0027 mm, subtending an angle of 1.3 arc seconds in object space. It would probably be concluded that the increased cost and complexity of the air-spaced configuration is not justified by this rather small decrease in spot size.

This is a good point in the lens selection procedure for considering the effects of diffraction on the performance of our lens. We have

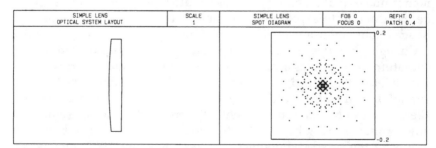

Figure 7-1. At left is a simple positive lens element with a 50 mm diameter and a 400 mm focal length. At right is a spot diagram generated by tracing 300 rays through that lens and plotting their intersections at the image plane. The large spot size is due to residual chromatic aberration.

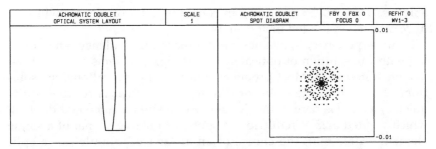

Figure 7-2. At left is a cemented achromatic doublet lens, with a 50 mm diameter and a 400 mm focal length. At right is a spot diagram generated by tracing 300 rays through that lens and plotting their intersections at the image plane. The resulting spot is due mainly to residual secondary chromatic aberration.

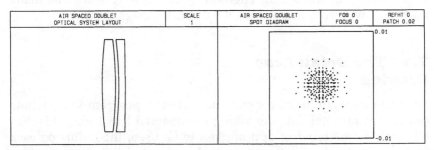

Figure 7-3. At left is an air-spaced achromatic doublet lens with a 50 mm diameter and a 400 mm focal length. At right is a spot diagram generated by tracing 300 rays through that lens and plotting their intersections at the image plane. The reduced spot size is essentially the same as that seen in Fig. 7-2.

established that this lens will function at the primary wavelength of 0.56 µm, the center of the visual spectrum, and that the speed of this lens is to be 400 mm EFL÷50 mm diameter = $f/8$. In a previous chapter it was stated that, assuming a complete absence of residual aberrations, a lens will still produce a finite blur spot due to diffraction effects at its aperture. The formula for the radius of that blur spot due to diffraction (the Airy disk) is

$$R = 1.22 \times \lambda \times f \text{ number}$$

for this lens:

$$R = 1.22 \times 0.00056 \times 8 = 0.0055 \text{ mm}$$

Since geometric analysis of both the cemented and air-spaced versions of the achromatic doublet indicate blur spots that are considerably less than 0.0055 mm, we can conclude that both lenses will produce diffraction limited image quality.

In my experience, I have observed a recurring tendency when dealing with the solution of optical system design problems that has been demonstrated here. Most problems will present several obvious solutions. Most often the point of diminishing return is reached quite quickly, as it was here in the decision to consider a cemented doublet, which yielded a 25 × reduction in spot size relative to that of a single element. After that, the choice of an air-spaced doublet, or a split crown-doublet, is found to yield improvement in performance that is essentially negligible, while increasing the cost significantly. Fortunately, today's computer analysis tools and methods, make it quite easy (and worthwhile) to explore these alternate designs. It is the responsibility of the optical engineer to carefully consider the many tradeoffs involved when making the ultimate design decision.

7.4 The Detail Lens Drawing

The computer output of a typical lens design program will include complete numerical data describing the nominal lens form. This data will include the types of optical glass to be used, the radius on each surface, the thickness of each lens element, and the clear aperture requirement on each surface. It is an important function of the optical engineer to convert this raw numerical data into a detail drawing, suitable for use by the optical shop in the manufacture of this lens. The detail drawing must include all nominal values, along with appropriate tolerances. Certain physical characteristics, such as lens diameter, will be determined by the mounting method and other mechanical design considerations. In this section we will examine the process of converting the lens design data for the 400-mm cemented doublet to the detail drawing format.

Figure 7-4 shows the computer printout for the cemented doublet. While all nominal data are present, this format must be modified and tolerances must be added in order to produce a document that will be usable by the optical manufacturing operation. Figure 7-5 is a finished working drawing of the same cemented doublet. While several different drawing formats are presently used throughout the industry, the generic format shown here contains all the information that is needed for the manufacture of a satisfactory lens. A great deal of work is presently under way, directed toward the generation of internationally common and acceptable optical manufacturing and drawing stan-

```
*PARAXIAL CONSTANTS
    EFL        FNB       GIH       PIV        PTZRAD       TMAG
 399.99933   7.99999   0.06981   -0.00436   -560.39617   -5.0000E-19

*LENS DATA
ACHROMATIC DOUBLET
SRF    RADIUS      THICKNESS   APERTURE RADIUS   GLASS   SPE   NOTES
 1      --            --        25.00000 S        AIR
 2    195.20000 V   8.00000     27.00000 A        BK7   C
 3   -177.50000 V   6.00000     27.00000          F7    C
 4  -1344.30000    391.43000    27.00000          AIR

 5      --            --         0.07210 S         AIR

*GENERAL DATA
OSLO 2.3  WLKA
   EPR          OBY         THO        CVO    CCO        UNITS
25.00000  -1.3963E+17  8.0000E+20      --     --        1.00000
  IMS     AST    RFS    AFO      AMO    DESIGNER    IDNBR
   5       2      2      0       TRA      OSLO       483

*WAVELENGTHS
CURRENT  WV1/WW1   WV2/WW2   WV3/WW3   WV4/WW4   WV5/WW5   WV6/WW6
  1      0.56000   0.50000   0.62000   0.43584   0.70652   0.40466
         1.00000   0.50000   0.50000     --        --        --

*REFRACTIVE INDICES
SRF   GLASS   RN1/RN4   RN2/RN5   RN3/RN6    VNBR
 1     AIR      --        --        --        --
 2     BK7    1.51803   1.52141   1.51554   88.17003
 3     F7     1.62802   1.63558   1.62271   48.79468
 4     AIR      --        --        --        --
 5     AIR      --        --        --        --
```

Figure 7-4. Computer output from OSLO Series 2, describing the nominal characteristics of the 400-mm, $f/8.0$ cemented achromatic doublet.

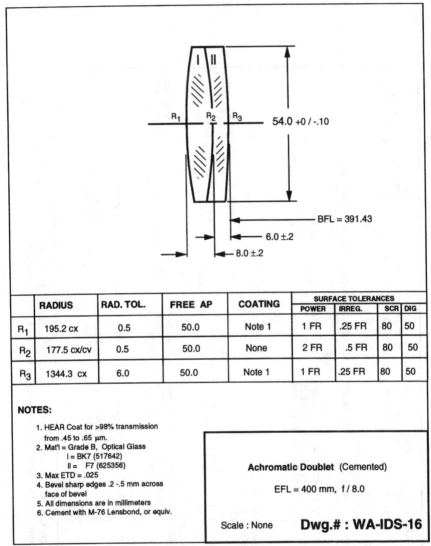

| | RADIUS | RAD. TOL. | FREE AP | COATING | SURFACE TOLERANCES | | | |
					POWER	IRREG.	SCR	DIG
R_1	195.2 cx	0.5	50.0	Note 1	1 FR	.25 FR	80	50
R_2	177.5 cx/cv	0.5	50.0	None	2 FR	.5 FR	80	50
R_3	1344.3 cx	6.0	50.0	Note 1	1 FR	.25 FR	80	50

NOTES:

1. HEAR Coat for >98% transmission from .45 to .65 μm.
2. Mat'l = Grade B, Optical Glass
   I = BK7 (517642)
   II = F7 (625356)
3. Max ETD = .025
4. Bevel sharp edges .2 -.5 mm across face of bevel
5. All dimensions are in millimeters
6. Cement with M-76 Lensbond, or equiv.

Achromatic Doublet (Cemented)

EFL = 400 mm, f / 8.0

Scale : None   **Dwg.# : WA-IDS-16**

Figure 7-5. A complete detail lens drawing of the cemented achromatic doublet.

dards. The ultimate selection of a drawing format will depend on the particular requirements of the customer and, to a lesser degree, those of the optical shop that is selected to produce the part.

The basic lens illustration should be drawn to scale, or at least should be an accurate representation of the actual shapes and proportions of the lens elements. This part of the lens drawing should contain all basic physical dimensions, such as lens thickness and diameter. Each optical surface is designated by a number (R_1, R_2, or R_3) and all relevant optical information for these surfaces is presented in tabular form.

The subject of tolerance analysis is critical to the success of the manufacturing stage. The approach taken will vary, depending on the complexity of the design and the analytical tools available. In this particular example it is a relatively simple matter to introduce small changes to the various lens parameters, determine the effect of these changes on image quality and then settle on an acceptable value for each tolerance. The tolerances established in this case were on radius, thickness, and element wedge, or edge thickness difference (ETD in Fig. 7-5, note 3). Other tolerances include surface sphericity, which is determined on the basis of the fact that, in this case, we are expecting essentially diffraction-limited performance from the finished lens. The cosmetic quality of each surface is covered by the "scratch and dig" specifications. In the case of an objective lens such as this, the relatively loose 80–50 cosmetic specification is acceptable. Were the lens located closer to an image plane, or were it for a laser application, rather than a visual system, then this scratch-and-dig tolerance would have to be reexamined and probably made tighter. The military standard on optics (MIL-O-13830) should be read and carefully considered by all involved in this area of optical tolerances and quality.

The drawing also contains a set of notes that cover the types and quality of optical glass to be used, the antireflection coatings and the cementing process. The free-aperture radius is given for each optical surface. This is important because optical and cosmetic quality need not be checked outside the specified free aperture. Perhaps the most important rule that should be applied to the preparation of optical component drawings is that they must be easily understood and completely free of ambiguity. If any of the items covered on the drawing is left open to interpretation, the drawing has not done its job and the resulting part will, most likely, not perform up to expectations. To state one final and obvious point, each drawing must have a unique drawing number that will allow it to be easily and accurately identified without confusion as work on the part progresses.

7.5 The Mirror

Any optical component whose function is to reflect incident light is called a *mirror*. In the design of a mirror, two factors are of special importance: the shape of the reflecting surface and the reflectivity characteristics of that surface. The simplest and most common mirror is the flat mirror. This component serves to deflect the optical path, while having no effect on the reflected light in terms of convergence, divergence, or optical aberrations. Because of this, the optical system can be designed ignoring any flat mirrors that may be introduced to the design at a later stage. One key factor that cannot be ignored is the effect that a mirror, or set of mirrors, will have on the orientation of the image produced by the optical system. While a precise and thorough evaluation must be made in all cases, a general rule of thumb for system design is that an odd number of mirrors within a system will result in a reversed image that will not be satisfactory. On the other hand, an even number of mirrors will generally leave the final image in an orientation that will be satisfactory.

To demonstrate this phenomenon, let's consider once again the 10 × visual telescope that was introduced during the discussion of simple lenses. Assume for the moment that we want that telescope to contain a 90° bend, such that in order to look straight ahead, we would actually be required to look straight down into the telescope. It can be seen from Fig. 7-6 that the addition of a flat mirror between the lens and its image will produce the desired 90° bend, but the resulting image will be reversed from left to right, which would not be acceptable in all cases. For example, if the telescope were intended for astronomical viewing, such an image reversal would not be considered a problem. However, were the telescope intended for viewing terrestrial objects, such as birds, animals, or sporting events, the reversed image would be clearly unacceptable. The most common solution to the image reversal problem is to use a "roof" mirror in place of the flat mirror (see Fig. 7-7). Critical to the performance of a roof mirror is the fact that the two reflecting surfaces be at precise right angles to each other. Any error in the 90° angle will result in the formation of a double image, which will adversely effect the image quality of the instrument. Typically, a tolerance of 1 to 5 arc seconds will be required on the 90° angle of the roof mirror assembly. Because of this tight angular requirement, the roof mirror is frequently replaced by a roof prism where the 90° angle can be more easily accomplished and is not subject to assembly and alignment problems (more on roof prisms later in this chapter).

The specification of optical quality on a mirror is related primarily to the shape of the surface and the type of reflecting material required. The mirror shown in Fig. 7-6 would be typical of folding mirrors found in any number of optical instruments. In this case the surface must be flat and since, as was stated earlier, the objective lens performance is to be diffraction-limited, the mirror should not introduce an error to the converging wavefront that is greater than $\frac{1}{8}$ wave. In

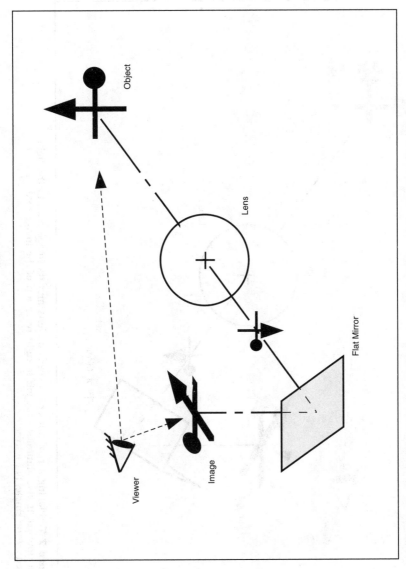

Figure 7-6. Placing a single flat mirror between the lens and its image reflects the light path upward. It also introduces a left-to-right reversion of the image as it appears to the viewer. In many applications this is an unsatisfactory condition.

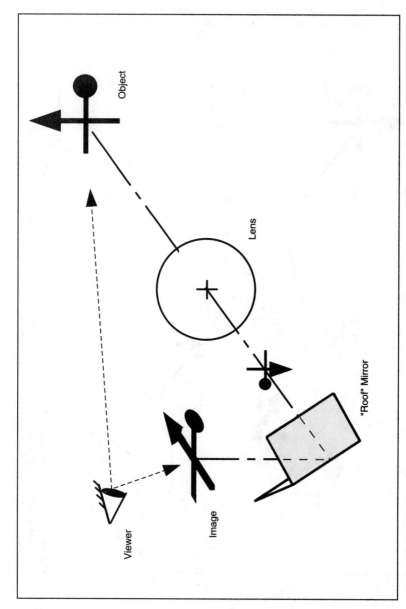

Figure 7-7. Placing a roof mirror between the lens and its image reflects the light path upward. It also eliminates the left-to-right reversion of the image that was produced by the single, flat mirror.

order to determine the actual aperture requirement for this mirror we must know the field of view of the lens and the location of the mirror relative to the lens and its image. In this case, the field of view is 3° and the mirror is to be located 300 mm from the back of the lens. With this information input to the lens analysis program, the ray trace shown in Fig. 7-8 can be generated. This shows us that the individual light bundle from each object point forms a 13 × 18-mm elliptical spot on the mirror surface and that the total clear aperture required on the mirror surface, to cover the full field of view, is a 30 × 42-mm ellipse.

It would be reasonable to specify that the mirror surface be flat to less than 0.1 wave over that 30 × 42-mm aperture. It will be of some value to consider how a tolerance of this magnitude might affect the reflected wavefront. Assume that the 0.1 wave irregularity appears in the form of a cylindrical mirror surface, that is perfectly flat in the cross section shown in Fig. 7-8, but curved in the section at a right angle (in and out of the page). The wavefront from any one object point will cover a 13 × 18-mm elliptical spot, as was described earlier. The departure of the mirror from flat across the 13-mm dimension will be a fraction of the total 0.1 wave that is allowed across the full 30-mm dimension. If the mirror error is 0.1 wave over 30 mm, it will be $(13 \div 30)^2 \times 0.1 = 0.019$ wave across the 13-mm dimension. Because this is a reflecting surface at a 45° angle, the resulting reflected wavefront error will be equal to 2.8 × the surface error, or 2.8 × 0.019 = 0.05 wave. Any error less than $\frac{1}{8}$ (0.125) wave will be consistent with our established goal of diffraction-limited performance for the telescope.

In selecting a reflecting material the major considerations are percent reflectivity, the wavelengths to be reflected, durability, and cost. In this case, where we are dealing with a conventional viewing device, a glass substrate with a deposited coating of protected aluminum, will generally be satisfactory. Such a coating will reflect about 85 percent of light within the visible spectrum, and it will also be quite durable and cost-effective. Figure 7-9 is a typical detail drawing of the mirror designed for this application. Making the overall shape rectangular (rather than elliptical) results in a piece that is more easily produced, while at the same time providing space for mounting hardware outside the clear aperture. Speaking of mounting, it is a common error to produce a mirror of high quality (flatness) and then to destroy the performance of that mirror by mounting it such that the surface is mechanically distorted. Extreme care must be taken in the mounting of all optical components, especially mirrors and prisms, to ensure that no mechanical stresses are introduced that might disturb the quality (flatness) of the reflecting surface.

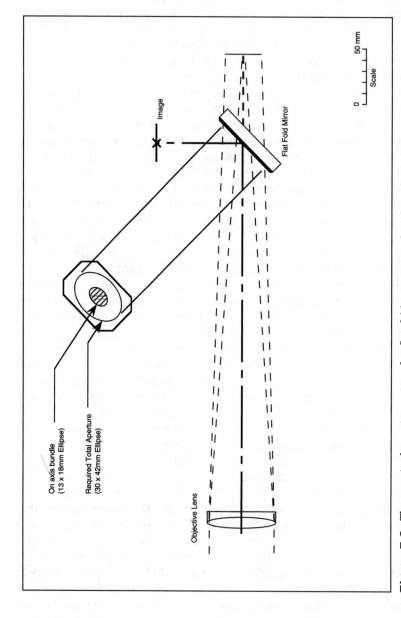

On axis bundle
(13 x 18mm Ellipse)

Required Total Aperture
(30 x 42mm Ellipse)

Objective Lens

Image

Flat Fold Mirror

0   50 mm

Scale

Figure 7-8. The required aperture size of a flat folding mirror placed in the back focus of a lens will depend on its axial location and the aperture and field of view of the lens.

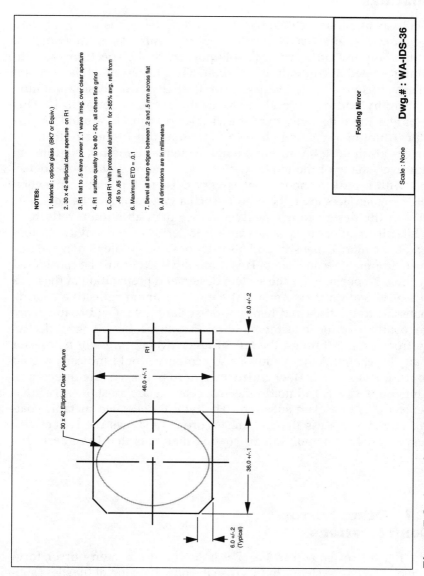

NOTES:

1. Material : optical glass (BK7 or Equiv.)

2. 30 x 42 elliptical clear aperture on R1

3. R1 flat to .5 wave power x .1 wave irreg. over clear aperture

4. R1 surface quality to be 80 - 50, all others fine grind

5. Coat R1 with protected aluminum for >85% avg. refl. from .45 to .65 μm

6. Maximum ETD = 0.1

7. Bevel all sharp edges between .2 and .5 mm across flat

8. All dimensions are in millimeters

30 x 42 Elliptical Clear Aperture

R1

46.0 +/-.1

8.0 +/-.2

36.0 +/-.1

6.0 +/-.2 (Typical)

Folding Mirror

Scale : None   **Dwg.# : WA-IDS-36**

Figure 7-9. A complete detail drawing of a front surface, flat folding mirror.

7.6 Common Reflective Coatings

The basic aluminum coating with a protective layer is the most common and cost-effective choice for reflecting surfaces in most optical system applications. In the visible portion of the spectrum, an aluminum-coated mirror will reflect about 85 to 90 percent of the incident energy. It is possible to increase the reflectivity of the basic aluminum coating by adding several layers of dielectric material to it. This improves the reflectivity over the visible spectrum to around 95 percent. Aluminum coatings have an inherent dip in their reflectivity curves at about 0.8 μm. For visual systems this is not a problem. However, many of the modern electronic imaging systems use detectors with a peak response that is very close to this value. If several reflecting surfaces are to be used in such a system, it may be advisable to shift the design to use a silver rather than aluminum reflectors. While silver offers a clear advantage in terms of reflectivity, it does tend to be more expensive and also is more susceptible to damage due to environmental conditions. This durability factor must be considered by those responsible for the system design and preparation of the coating specification. Last among the more common reflective coating materials to be discussed here is first-surface gold. Gold coatings are frequently used for reflectors in infrared systems. Its average reflectivity from 0.8 μm through the far infrared (IR) (12 μm) will be greater than 96 percent. As its yellowish appearance would indicate, a gold coating does not reflect uniformly across the visible spectrum. Aluminum, silver, and gold reflective coatings are most often vacuum deposited onto a glass substrate. The typical deposited metallic coating thickness is less than 1 μm. Figure 7-10 shows the reflectivity curves for the common mirror coating materials that have been discussed here.

7.7 Other Mirror Configurations

The flat mirror shown in Fig. 7-9 is the most common mirror form found in optical systems. In this case the only function of the glass substrate is to provide a stable platform for the reflective coating of vacuum-deposited aluminum. Because the light reflects directly from the aluminum, never entering the glass, this is referred to as a *front surface*

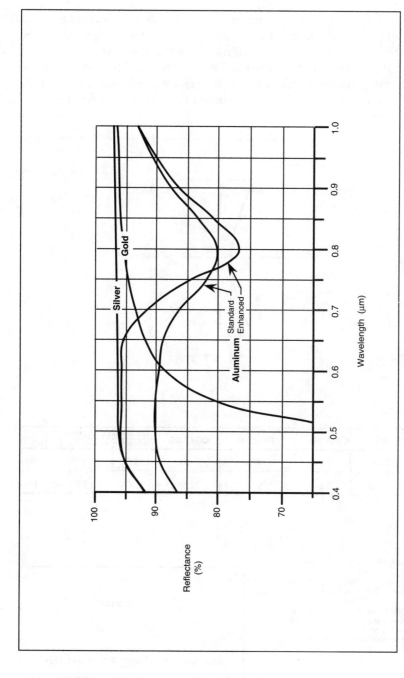

Figure 7-10. A number of materials are used for the reflecting coating on optical mirrors. Shown here are the reflectance curves for the most commonly used materials.

mirror. While a rear surface mirror offers the obvious advantage of a very well protected coating, the negative impact of passing through the glass substrate usually far outweighs that advantage. An exception to this would be the Mangin mirror form shown in Fig. 7-11. Here, a curved, rear surface mirror and weak lens are combined into a single element for purposes of aberration correction. This mirror form is often

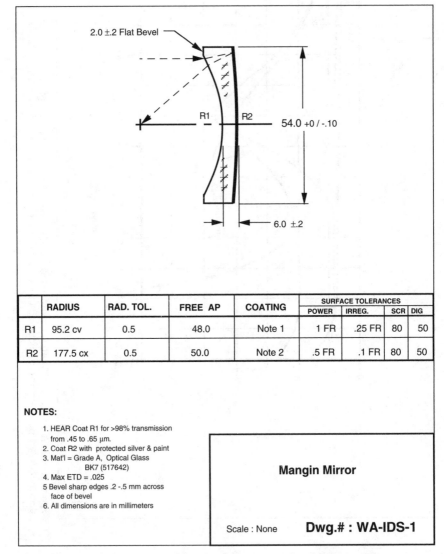

	RADIUS	**RAD. TOL.**	**FREE AP**	**COATING**	**SURFACE TOLERANCES**			
					POWER	**IRREG.**	**SCR**	**DIG**
R1	95.2 cv	0.5	48.0	Note 1	1 FR	.25 FR	80	50
R2	177.5 cx	0.5	50.0	Note 2	.5 FR	.1 FR	80	50

NOTES:

1. HEAR Coat R1 for >98% transmission from .45 to .65 μm.
2. Coat R2 with protected silver & paint
3. Mat'l = Grade A, Optical Glass BK7 (517642)
4. Max ETD = .025
5 Bevel sharp edges .2 -.5 mm across face of bevel
6. All dimensions are in millimeters

Mangin Mirror

Scale : None   **Dwg.# : WA-IDS-1**

Figure 7-11. A complete detail drawing of a rear surfaced Mangin mirror.

used as the primary reflector in the design of astronomical telescopes.

Until recent years the polished glass substrate with a deposited reflective coating was the mirror configuration of choice for use in most optical systems. Recently, two new processes have been developed which offer alternative approaches to the optical designer. They are replicated and micromachined mirrors. In both cases the principle advantage is that the mechanical means for mounting the mirror can be made an integral part of the mirror itself.

In the case of the replicated mirror, there is also a considerable cost advantage when large quantities are involved. For the production of a flat, first-surface mirror, the replication process begins with a master flat that has been polished to the precision required of the final mirror. As is shown in Fig. 7-12, the first step in the process is to cover the master with a parting layer of a *nonstick* material such as silicone, that will not adhere to adjacent layers. This is followed by the vacuum deposition of the required reflective coating, with the layers deposited in reverse order. Meanwhile, the mechanical substrate surface has been machined flat to normal mechanical tolerances and the part is stress-relieved as required. A thin layer of epoxy is then applied to the flat surface of the substrate, and the epoxy is brought into intimate contact with the prepared master. After the epoxy has cured, the finished part may be separated from the master, with the coated surface now becoming an integral part of the mechanical substrate. The process derives its name from the fact that the final reflecting surface is a precise replica of the polished master. Depending on the sizes involved, it may be possible to produce several replicated mirrors simultaneously, using the same master.

Micromachining, or *diamond point machining,* is a very precise method of machining metal to tolerances that are consistent with the requirements of IR optical components. This distinction arises because the surface accuracy required will be a function of the wavelength involved and most IR systems operate at a wavelength that is 10 to 20 times larger than that of visual systems. While micromachined surfaces must be produced individually, it is possible to produce them using computer-controlled machines that are essentially automatic and very precise. An additional major advantage of the micromachined mirror is that the machine can be programed to produce a nonspherical (aspheric) surface as easily as a sphere. The introduction of asphercs makes possible the design of better-performing systems with fewer elements, which results in substantially reduced cost. In addition to mirrors, there are several IR refracting materials that can also be micromachined. This makes it possible to produce aspheric IR lenses as well as mirrors using the micromachining process.

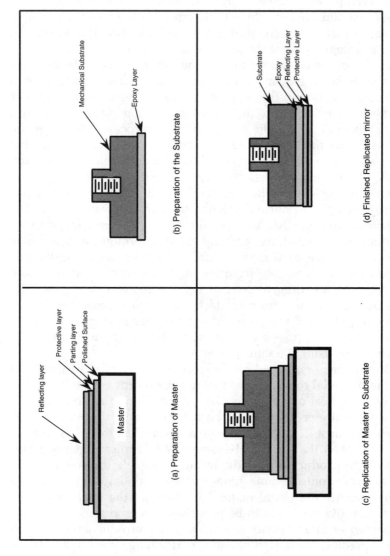

Figure 7.12. A relatively recent development in the field of optical manufacturing makes it possible to replicate the polished surface of a master onto a mechanical substrate.

The micromachining process has also been used to produce mirrors for visual optical systems. The very fine residual machining pattern left on the surface by the process has a tendency to cause problems with scattered light and diffraction in systems operating at visible wavelengths. While it is possible to eliminate the machine pattern by postpolishing of the micromachined surface, this approach eliminates many of the advantages offered by the micromachining process.

7.8 Prisms

Like the mirror, the *prism* is an optical component used most often to deviate or displace the optical axis. Perhaps the most common prism form is the *right-angle prism* that would typically be used to replace a flat mirror such as was shown in Fig. 7-6. There are several advantages to using a prism in this application. First, the prism will be easier to mount because it is a much more rigid component than the mirror. Second, in the prism the reflective coating surface will be protected from physical and environmental damage. Once again, these advantages do not come without some tradeoffs. The prism must be made from glass that is of high optical quality, while the substrate quality of the mirror is relatively unimportant. To produce the right-angle prism, three of its surfaces must be polished flat, where the mirror requires just one. Finally, the prism will have an effect on image quality when it is used in a converging light bundle, as is the case in this example.

We will examine in some detail the effect that the addition of a right-angle prism will have on the image quality of our previously selected cemented doublet lens. It will be recalled that this lens has a focal length of 400 mm, and a diameter of 50 mm and it produced a blur spot with a radius of 0.003 mm, due primarily to residual secondary color. From Fig. 7-8 it can be determined that a 40-mm right-angle prism can be used to replace the flat mirror, with a lens to prism distance that will be about 280 mm. Two significant changes will be seen at the image plane due to the insertion of this prism. First, the basic axial location of the image will be shifted away from the lens by the following amount:

$$\text{Focus shift} = (n - 1)\frac{t}{n}$$

where n is the prism index of refraction and t is prism thickness.

In this case the index is 1.52 and the thickness is 40 mm. The resulting focus shift will be about 13.7 mm. A useful rule of thumb, applicable to most prisms made from glass with an index of about 1.5, is that the image will be shifted by approximately one-third the prism thickness. The effect on image quality is easily determined by adding the 40 mm of glass to our computer model of the lens and evaluating the resulting lens performance. In this case it is found that the spherical aberration is increased and the correction of primary color is slightly disturbed. However, execution of the spot diagram analysis indicates that the end result is to increase the spot size from its original value of 0.003 to 0.0037 mm. This is still well within the diffraction limit for this lens, which we found to be 0.0055 mm, so we may conclude that the changes due to adding the prism will be negligible. The computer analysis also permits us to confirm the 13.7-mm focus shift due to insertion of the prism that we had calculated earlier.

Insertion of a prism into the back focus of a lens is not always this simple. For example, a similar analysis shows that making the prism thickness 100 mm in this case would increase the spot size such that it was equal to the diffraction limit. In that case it would probably be advisable to modify the lens prescription to a form where the spherical aberration and primary color have been recorrected. A typical right-angle prism is shown in Fig. 7-13a.

As was the case with the flat mirror, the right-angle prism, with its single reflecting surface, will cause a reversal of the image that is unacceptable in certain applications. In this case it would be appropriate to use an equivalent of the roof mirror, which would be a right-angle prism with a roof surface in place of the flat reflecting surface. A prism of this type is traditionally referred to as an *Amici prism* (see Fig. 7-13b). Because of the roof surface, the total path through the prism will be about 1.7 × the aperture diameter. In our example (with a 40-mm diameter bundle) this would mean the glass path would be 68 mm and the focus shift, about 23 mm.

Figure 7-13c–g illustrate several other common prism configurations. The *Porro prism* will be found in traditional prism binoculars. By introducing a pair of Porro prisms into the optical path, the image from the objective lens is brought to the correct orientation for straight ahead viewing through the eyepiece. The *dove prism* is often used as a derotation prism within an optical instrument. Many instruments are designed to have a stationary eyepiece and viewing point, while a scanning mirror directs the line of sight in various directions (see Fig. 7-14). As the illustration shows, as the instrument scans 360° in azimuth, the image being viewed will appear to rotate through that same angle, about the optical axis. A dove prism can be linked to the scan mechanism in such a way as to produce a counter rotation that will offset that problem, resulting in a rotation-free image. The *delta*

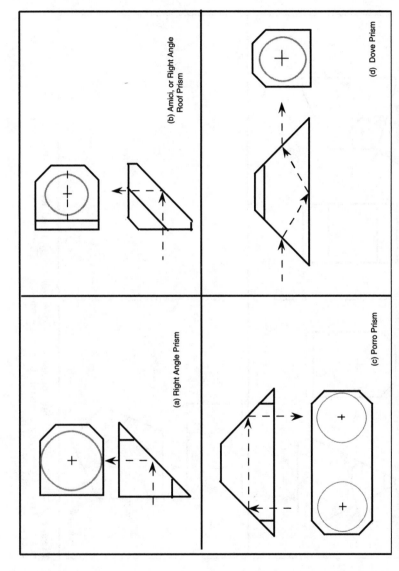

Figure 7-13. Various prism types are commonly used in optical system design. Prisms generally function to change the direction of the optical axis or to modify the orientation of the final image.

(a) Right Angle Prism

(b) Amici, or Right Angle Roof Prism

(c) Porro Prism

(d) Dove Prism

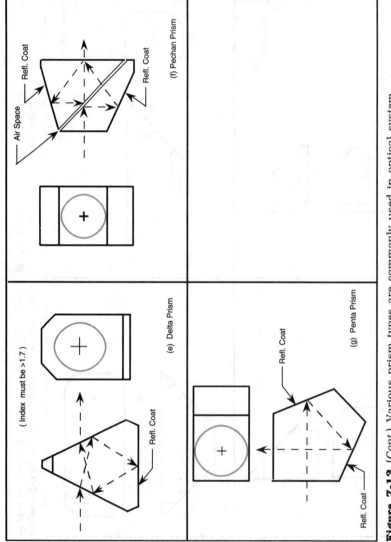

Figure 7-13 (*Cont.*) Various prism types are commonly used in optical system design. Prisms generally function to change the direction of the optical axis or to modify the orientation of the final image.

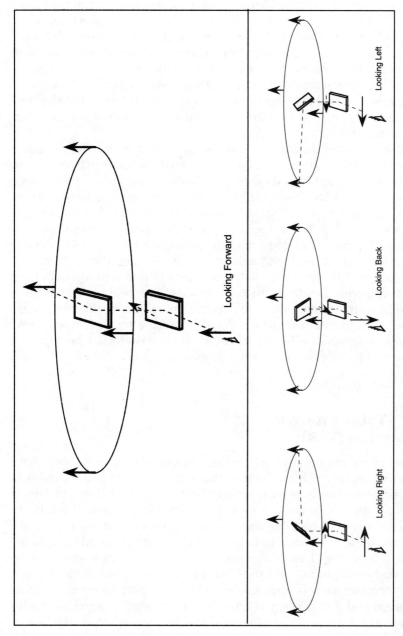

Figure 7-14. Diagram illustrating the rotation of the output image created by an azimuth scan using the top mirror in a two-mirror (golf course) periscope viewer.

prism is a very compact (folded) version of the dove prism that must be made from a relatively high-index glass in order to work properly. It should be noted that in the cases of the dove and delta prisms the optical axis strikes the first prism surface at a considerable angle of incidence. Because of this, it is not possible to use either of these prisms in a converging light bundle. The incident light bundle must be collimated, such that it appears to come from infinity. On the plus side, the prism placed in collimated light will not introduce any aberrations into the system.

A third form of derotation prism is the *Pechan prism.* The Pechan is made up of two separate prisms with a small airspace between them. In this case, the optical axis is perpendicular to the entrance face. As a result, the Pechan prism can be used in converging light, but its impact on image quality must be taken into account. The final prism form shown in Fig. 7-13 is the *penta prism.* Its name is derived from the fact that this prism has five sides. The penta prism functions to deviate the optical axis through an angle of exactly 90°. Because of its unique construction, this prism is relatively insensitive to rotation about the two axes perpendicular to the optical axis. The penta prism is often found in instruments such as the optical range finder, where a precise 90° angular deviation is critical. As was the case for the Amici prism, the reflecting surface in any of these other prisms may be replaced with a roof in order to achieve the desired image orientation.

7.9 Total Internal Reflection (TIR)

Several of the prisms covered here will function without the need for a reflective coating. The reflection that occurs within these prisms is referred to as *total internal reflection* (TIR). It will be helpful to review this topic and to establish a basic understanding of how it works. In Fig. 7-15 we see a block of glass with two faces, labeled A and B. In the first case these faces are parallel to each other and perpendicular to the incident light ray. This is nothing more than a common optical window, and the light ray will pass through it undeviated. It can be seen that when surface B is tipped relative to A, the part becomes a deviating prism and the light ray is refracted at surface B in agreement with the law of refraction. When the condition is reached where the sine of the angle of refraction is ≥ 1.0, then there will be no refraction, and all the energy will be reflected internally. This angle of incidence is

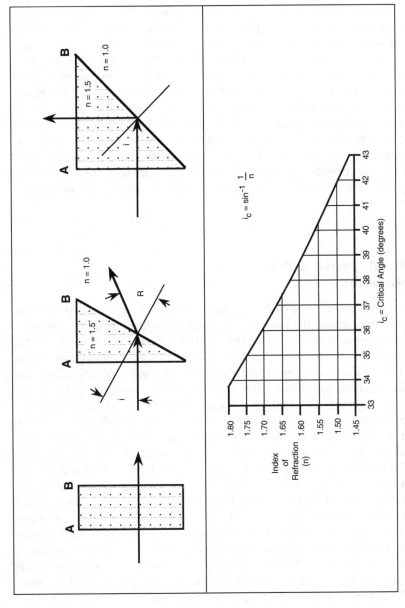

Figure 7-15. Total internal reflection (TIR) occurs when the sine of the angle of incidence within the glass is greater than 1/ the index of refraction. This is referred to as the *critical angle.*

called the *critical angle*. TIR is a very simple and 100 percent efficient phenomenon. When it works, it works very well. However, there are a few dangers related to its application that should be considered. First, the TIR surface must be kept free of contaminants such as water or fingerprints that would interfere with TIR. For that reason, the TIR surface cannot be used as a mounting surface. Also, the application must be carefully analyzed to ensure that none of the system's light rays are incident at less than the critical angle. Quite often the case will occur where the optical axis meets the TIR criteria, but other light rays within the system do not. For these, and other reasons, it is often a reasonable approach to apply a reflective coating to the surface, accept the 5 to 10 percent reflectivity loss, and avoid all of these potential trouble spots. The graph in Fig. 7-15 plots the critical angle as a function of index of refraction.

7.10 A Typical Prism Application

Whenever a prism is added to an optical system, it is crucial that the designer very carefully analyze all the factors that might be involved. Arguably the best reference on this subject is the prism section of *MIL Handbook 141*. A thorough understanding of that section, combined with the application of modern computer design tools, will allow the optical engineer to successfully incorporate the required prisms into the optical system. For example, assume that we wish to add a derotation prism in the back focal region of our 400-mm doublet. At the same time it is important that the final design be as compact as possible. The logical selection to meet these requirements would be a Pechan prism. The data sheet contained in *MIL Handbook 141* tells us that the Pechan prism may be made of BK7 optical glass (an ideal prism material), its internal length or light path is 4.62 times its aperture size, and its physical length from entrance to exit face is just 1.21 times the aperture size. Knowing that the prism will be quite close to the objective lens, and that the lens has an aperture of 50 mm, we may safely assume a prism that also has a 50-mm aperture. The constants from the handbook tell us that the total glass thickness will be 4.62 × 50 = 231 mm, and the actual (mechanical) prism thickness will be 60.5 mm. The next step in the design process would be to insert a 231-mm-thick glass block into the back focus of our previously designed 400-mm cemented doublet. As we would expect from our earlier analysis, the correction of the lens aberrations will be seriously degraded. What was previously a 0.003-mm spot radius has been increased to 0.010 mm because of increased spherical and chromatic aberration. This spot size is twice the diffraction limit, which is our performance goal for this lens. As a result, it will be necessary to reoptimize the lens, to compensate for the aberrations of the Pechan prism. A positive feature of this prism is that, because the optical axis is essentially folded around itself, the physical length of the prism is substantially less than the actual optical path. In this case the physical length is about 60 mm, while the optical path is 231 mm. We have learned that the insertion of 231 mm of glass into the back focus of a lens will increase the distance to the image plane by about one-third the glass thickness, or 77 mm in this case. On the other hand, the Pechan prism has a physical length of just 60 mm, 171 mm less than the actual optical path within the glass. The net result is that, by inserting the Pechan prism into our lens system, we will reduce the overall length from lens to image by 171 − 77 = 94 mm. Figure 7-16 shows the optical system with and without the Pechan prism in place. In addition to the image derotation capability that has been added, a 94-mm reduction in overall physical size has been realized.

7.11 Review and Summary

This chapter has been intended to give the reader a feel for some of the basic optical components commonly found in today's optical systems. These component types have been identified as lenses, mirrors, and prisms. The simple positive lens is the most frequently encountered optical component. Its various forms have been discussed and the relative performance of the single element, the cemented doublet, and the air-spaced doublet evaluated. A hypothetical lens application has been presented for purposes of demonstrating design procedures and the evaluation of relative performance. The preparation of detail drawings of optical components for use by the optical shop for manufacture has been covered in some detail.

The section on design and construction of mirrors for optical systems included the most common mirror type (first-surface flat) and also a few of the more complex forms. A brief discussion of replicated and micromachined mirrors was included. Finally, the topic of prisms has been covered, including illustrations of the most common prism types along with a discussion of their typical functions. A detailed

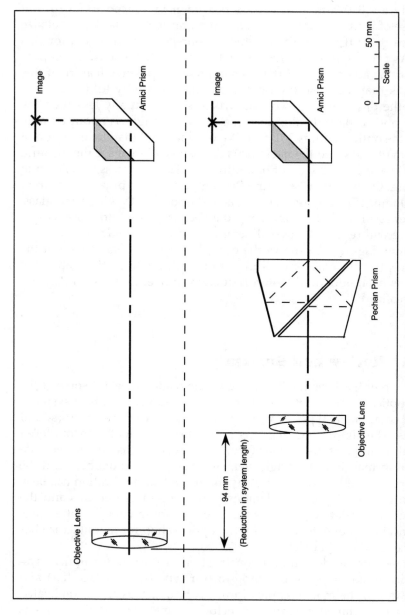

Figure 7-16. Flowchart illustrating the effect of adding a Pechan derotation prism to the back focus of a 400-mm doublet.

example of the impact that the addition of a prism has on overall system design has been presented.

All of this has been intended to make the reader familiar and comfortable with these very basic optical components. Hopefully, it will prepare the optical engineer to work with more advanced reference material and with modern optical design and analysis computer software, to further develop essential design skills in these areas.

8

Basic Optical Instruments

8.1 Introduction

An *optical instrument* can be defined as a device used for observing, measuring, and recording optical data or information. It is often the responsibility of the optical engineer to design the optical system to be incorporated into such a device. This chapter will describe several basic instruments and discuss their applications and some of the special aspects of these instruments, as they affect their design and manufacture. The most common optical instruments are those intended to enhance visual capacity. This chapter will deal primarily with instruments of that type.

8.2 The Magnifier or Loupe

A magnifying glass (or Loupe, from the Old French, meaning an imperfect gem), is the simplest of optical instruments intended for the enhancement of visual capability. It is a device frequently associated with jewelers, usually taking the form of a simple lens. In use, the magnifier is held close to the eye, while the object being viewed is placed near its image plane.

The average young and healthy human eye is capable of focusing down to a minimum distance of 254 mm (10 in). That same eye is able to resolve a repeating, high-contrast target of equal-width black and white lines, when each line subtends an angle of at least 1 arc minute ($\frac{1}{60}$th of a degree). Most often, when we view an object it is our intent to distinguish as much detail within that object as is possible. To that end we first bring the object as close as possible to the eye. When that closest distance is 254 mm, the smallest resolved detail on the object, which subtends an angle of one minute (tan = 0.0003), will have an actual size of $254 \times 0.0003 = 0.076$ mm. If this resolved element is part of a repeating pattern of equal-thickness black and white lines, then each cycle (one black plus one white line) will have a thickness of 0.152 mm. The frequency of this finest resolvable pattern then will be $1/0.152 = 6.6$ cycles/mm.

A *magnifier* is any positive lens with a focal length that is less than 254 mm. The amount of magnification M that a lens will provide is easily calculated by dividing its focal length into 254. For example, a 50-mm lens would provide a magnification of $M = 254 \div 50 = 5.1 \times$. This formula applies to the standard case, where the object is placed at the focal point of the magnifier and the image being viewed appears at infinity. This condition allows most comfortable viewing, with a relaxed eye. When the object is moved just a bit closer to the lens, such that the final image appears 254 mm from the eye, then the magnification will be increased by an additional $1 \times$.

A real-world example will be helpful in understanding how the simple magnifier can actually improve our ability to resolve detail. The back of a U.S. $5 bill contains a picture of the Lincoln Memorial. That structure is dominated by 12 columns across its facade. In the stonework above these columns, centered on the spaces between them, are the names of 11 states. The line thickness on these letters (on the bill) is approximately 0.02 mm, while the letter height is about 0.4 mm. When viewed at a distance of 254 mm, each letter subtends an angle of 5 minutes, while each line subtends an angle of just $\frac{1}{4}$ minute to the eye. Because this is considerably less than our visual resolution limit of 1 minute, it is not possible read the names of the states with the naked eye. When this portion of the bill is examined using a magnifier with a focal length of 50 mm, the eye's resolution is improved by a factor of $5 \times$, making it possible to read the names of the states. This experiment can be done using a removable 50-mm objective lens from a 35-mm single-lens reflex (SLR) camera to view a bill that is in reasonably good condition.

The subject of tradeoffs seems to keep coming up whenever we discuss various optical situations. In this case it is notable that when viewing with the naked eye we can easily determine that this is a U.S. $5 bill, containing a picture of the Lincoln Memorial. When viewed with a $5 \times$ magnifier, we can now see that the state name to the far left is Delaware, but the fact that this is a piece of currency becomes a matter of speculation. In optical terms, our field of view has been reduced from about 200° with the naked eye(s), to something around 40° with the magnifier. We have traded field of view for finer detail in the object being viewed.

Figure 8-1 contains ray trace analyses for four of the more common forms of simple magnifiers. Each has been analyzed for 10 × magnification (25-mm EFL), with a 4-mm-diameter pupil and a 30° field of view. A single element in both a symmetrical and planoconvex form is shown. The planoconvex lens is seen to have slightly more spherical aberration and less distortion. These differences are small enough to be considered negligible. Also shown are two achromatic forms of magnifier, a cemented doublet and a Hastings triplet. The obvious advantage of the triplet is its extremely well corrected on-axis performance. In all other regards it offers little or no advantage over the doublet. In most applications the achromatic doublet would probably be the preferred lens from among these four.

It is always a challenge, and should be the goal of the designer, to produce optical performance data that is complete, meaningful, and useful. In the case where we wish to select a magnifier type, it would be desirable to have some additional information regarding their relative image quality. Understanding the application, it can be concluded that the on-axis image quality is most important. Being used as a visual aide, it follows that only the central portion of the image will be critically examined, due to the limited high resolution field of view of the eye. In normal use, when examining an object, the tendency would be to bring whatever is critical into the central portion of the magnifier's field of view. Because of this, a more in depth analysis of on-axis image quality would be an appropriate tool for comparison of the various forms of magnifier. The first attempt at this might involve spot diagram analysis, which would yield a statistical spot radius for the four forms with the following results:

Form	RMS spot radius, mm
Planoconvex	0.015
Symmetrical	0.010
Doublet	0.007
Triplet	0.003

These results are useful in that they do give a good indication of how each form compares with the other. More useful would be an analysis that indicates the actual final performance of each magnifier when used with the eye. This is best accomplished by computing the

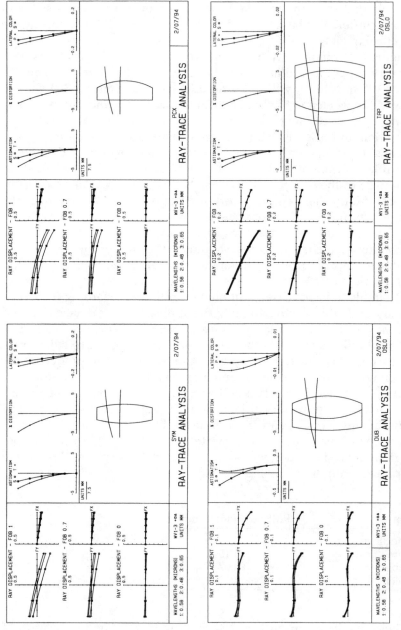

Figure 8-1. The simple magnifier, or Loupe, may take one of the four forms shown here. Ray trace analysis permits an estimate of their relative optical quality.

modulation transfer function (MTF) for each lens. The MTF curve indicates the modulation (contrast) of the image formed by the lens as a function of frequency. Figure 8-2a illustrates the concept of modulation and contrast, showing how they are measured and calculated. The modulation and contrast of the object (or image) is arrived at by comparing the maximum and minimum brightness values as follows:

$$\text{Modulation} = \frac{B_{max} - B_{min}}{B_{max} + B_{min}}$$

$$\text{Contrast} = \frac{B_{max} - B_{min}}{B_{max}}$$

For MTF analysis, the object being imaged is assumed to be a series of parallel black and white lines, of varying frequency, whose brightness varies as a sine function. The modulation of a high-contrast object is assumed to be 1.0 at all frequencies. In the image, when the frequency is very low, the modulation will be close to 1.0. As the frequency at the image increases, the modulation of the image will decrease, as a result of diffraction effects and aberrations of the lens. Figure 8-2b shows the on-axis MTF curves for the four magnifiers under consideration. Also shown is the diffraction limit curve, which represents the maximum MTF possible for a lens of this aperture, operating at this wavelength. Figure 8-2b also contains a dotted curve that is labeled *visual response curve*. This curve indicates, for a *typical* visual system using a 10× magnifier, the amount of modulation required in order to resolve the corresponding frequency. It should be noted that there is a great deal of data available relating to the performance of the human visual system, for a variety of subjects, under a great variety of circumstances. The curve shown in Fig. 8-2b is taken from one such set of data. It is *typical* performance, and as such it is appropriate for the intended purpose of comparing the quality of four similar designs. It should not be interpreted as an absolute indicator of visual performance under all circumstances.

The value of including the visual response curve is that its intersection with the MTF curve will indicate the resolution limit of the combination of the lens and the eye. We can conclude from Fig. 8-2b that, using a perfect (diffraction-limited) 10× magnifier, this particular combination would be able to resolve 78 cycles/mm at the object. Using the actual magnifiers, it can be seen that the triplet will resolve 77 cycles/mm (essentially perfect), the doublet 73 cycles/mm, the symmetrical single element 67 cycles/mm and finally, the planoconvex single element will resolve 58 cycles/mm. These hard and fast

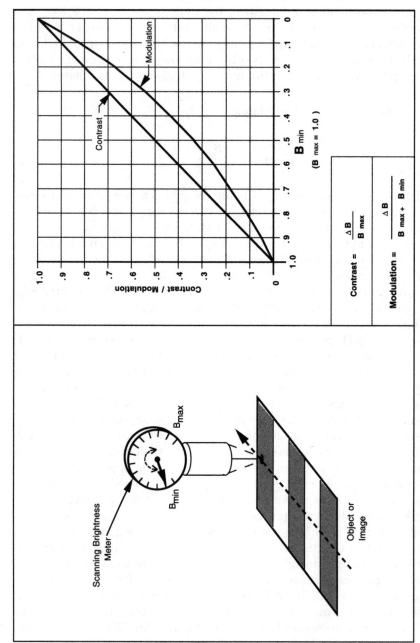

Figure 8-2a. Image quality is often described in terms of contrast, or modulation. This figure illustrates the definition of each and a method of measurement.

$$\text{Contrast} = \frac{\Delta B}{B_{max}}$$

$$\text{Modulation} = \frac{\Delta B}{B_{max} + B_{min}}$$

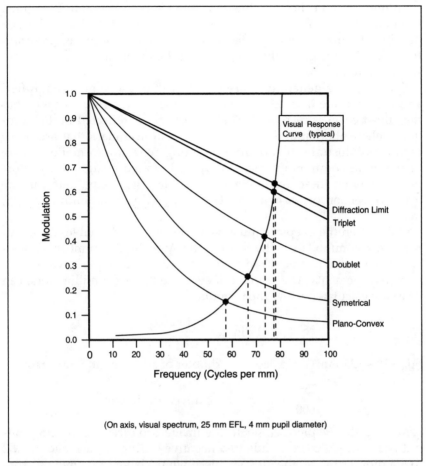

Figure 8-2b. Modulation transfer function (MTF) curves for four simple forms of magnifiers. Visual response curve permits an estimate to be made of actual resolution limit for the eye-magnifier combination.

results should allow the potential user to make an informed and intelligent selection from the four available designs.

8.3 The Eyepiece

The eyepiece is quite similar in function to the magnifier. It differs primarily in that it is generally used in conjunction with other optics to form a complete instrument, such as a telescope or microscope. In that type of application, the eyepiece serves two functions simultaneously:

(1) it must project the final image to the viewer's eye and (2) it must form an image of the system's aperture stop, which becomes the exit pupil of the instrument. While the first is by far the more important function, the design of the eyepiece requires that the second be paid some attention.

Eyepieces are generally more complex than the simple magnifier, and they become increasingly more complex as the field of view that they must cover is increased. That complexity is reflected in the number of elements required and the types of optical glass that are used. Figure 8-3 illustrates this point, presenting general information on six of the more common eyepiece designs. Detailed design and performance data on these and other eyepiece designs can be found in the eyepiece section of *Modern Lens Design,* by Warren J. Smith (App. A, item 7).

Quite often the eyepiece must be made axially adjustable, to permit focus to accommodate differences in the eyesight of a number of viewers. A normal adjustment range of from +2 to −4 diopters will satisfy most requirements. The amount of eyepiece travel (in millimeters) can be easily calculated using the formula

$$1 \text{ diopter} = \frac{\text{EFL}^2}{1000}$$

For a 10× (25-mm) eyepiece, each diopter of adjustment would require

$$\frac{25^2}{1000} = 0.625 \text{ mm of travel per diopter}$$

Moving the eyepiece closer to the image produces a diverging output beam which corresponds to a negative setting. To achieve the +2 to −4 diopter range, the 10× eyepiece would have to be moved from −2.5 to +1.25 mm, relative to the infinity, or zero diopter, setting. The eyepiece mechanism should be provided with a scale indicating the approximate eyepiece diopter setting.

8.4   The Microscope

A reasonable upper limit to magnification by a simple magnifier lens is about 20×. When greater magnification is required a compound form is generally used. This compound magnifier is commonly referred to as the *microscope.* In its simplest form the microscope consists of two lens assemblies: an objective lens and an eyepiece. The objective lens forms a magnified image of the object and the eyepiece

Type	Huygenian	Ramsden	Kellner
Layout	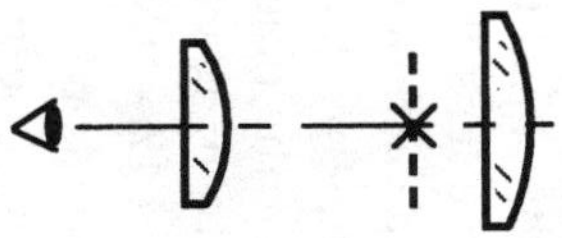	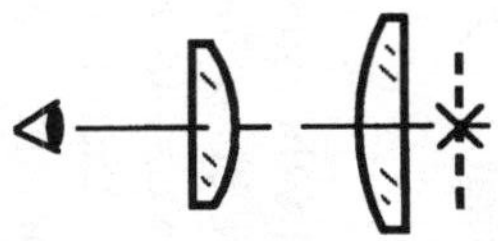	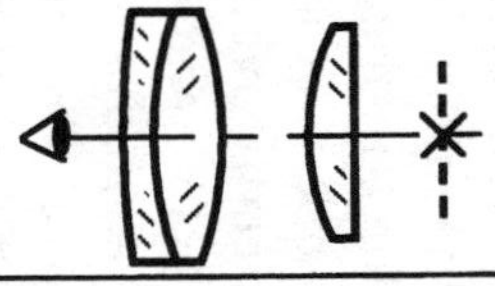
Field of View	25°	25°	35°
No. of Elements	2	2	3
Glass Types	BK7	BK7	BaK2, F4

Type	Orthoscopic	Erfle	Scidmore
Layout	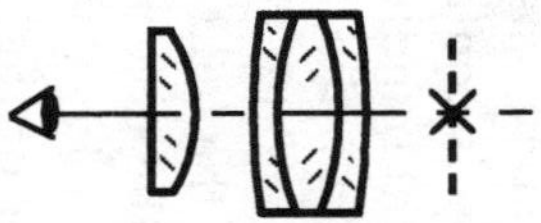	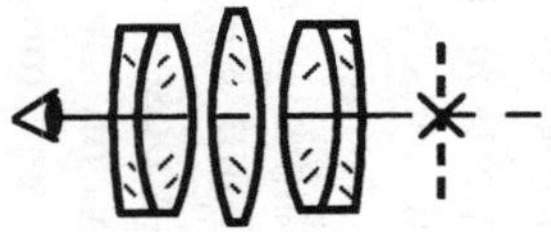	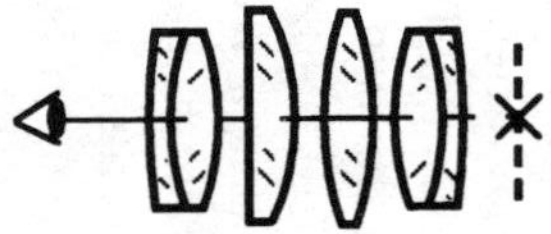
Field of View	40°	60°	80°
No. of Elements	4	5	6
Glass Types	BK7, KF3, F3	F2, BK7, SSK1 SK4, SF12	SF12, SK16

Figure 8-3. A family of common eyepiece designs, showing the increase in complexity as a function of field angle to be covered.

is used to view that image, providing additional magnification. In Fig. 8-4, we see a typical microscope arrangement, where the objective lens is used to form a 4× magnified image of the object and a 10× eyepiece is used to view that image. The total microscope magnification is found by multiplying the two individual magnifications, in this case yielding a total of $4 \times 10 = 40×$. Most often a microscope is designed by selecting from among a wide variety of commercially available components. While custom microscope design is an active and challenging field of optical engineering, it will not be covered in this book.

The information shown in Fig. 8-4 represents a typical, low-power microscope system, constructed from readily available commercial parts. The key to a successful design of this type is to identify a reliable source for these parts and to work closely with that supplier to assure that all assumptions you might make regarding optical and mechanical characteristics are valid.

In examining the details of Fig. 8-4, we see that both the objective and the eyepiece are quite similar to designs that have been covered earlier. For analysis purposes, the objective lens can be considered as imaging the field stop onto the object surface. The lens then takes on the familiar Petzval form, modified somewhat to function at a $\frac{1}{4} \times$ magnification rather than having the object at infinity. The field stop size, 18 mm diameter in this case, will limit the size of the object that is seen. An aperture stop is contained within the objective lens to limit the diameter of the light bundle that is transmitted from each point on the object. Vendor data indicate that the numerical aperture (NA) of this 4× objective lens is 0.12, producing an f number, at the object, of $1/(2 \text{ NA}) = f/4.17$. At the field stop that f number will be increased by the magnification of the objective lens, making it $4 \times 4.17 = f/16.68$. Since we are working with a 10× eyepiece, its focal length will be 25 mm and the diameter of the exit pupil for this microscope will be $25/16.68 = 1.52$ mm.

In a similar manner, we can analyze the microscope's field of view. The 18-mm-diameter field stop limits the actual field of view to $18 \div 4 = 4.50$ mm diameter at the object. Viewing the 18-mm-diameter image through the 25-mm eyepiece, we will have an apparent half field of view whose tangent is $9/25 = \tan^{-1} 0.36 = 20°$. The net result (ignoring eyepiece distortion) is that the 4.5-mm-diameter object, which would subtend about 1° to the naked eye when viewed at the 254-mm near point, will subtend an angle of about 40° when viewed through the microscope. This confirms the 40× magnification calculated earlier for this microscope assembly.

In all optical instruments the subject of focus must be given some attention. In the case of the microscope shown in Fig. 8-4, it would be

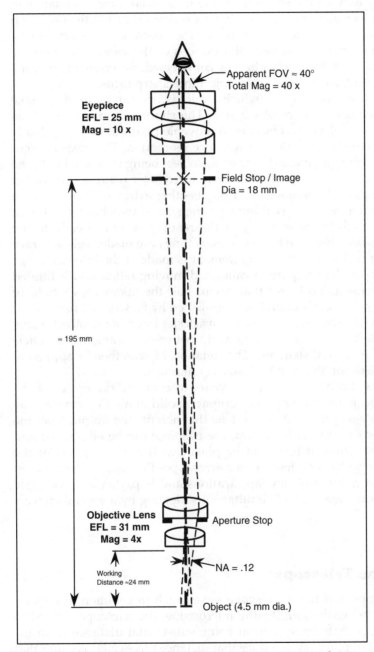

Figure 8-4. A typical compound microscope constructed using standard, commercially available components.

focused by moving the entire assembly in an axial direction relative to the object. In many cases there will be a reticle located at the field stop position within the microscope. It will then become necessary to first adjust the eyepiece to achieve sharp focus of the viewer's eye on the reticle pattern. After this has been accomplished, the entire microscope can then be adjusted to bring the object into sharp focus.

There are several ways in which a microscope can be used to make precise measurements of object size. The simplest and most obvious approach is to place the object on a movable platform (a stage) that is equipped with micrometers to measure its motion. The measurement is made by aligning one extreme of the object being measured with the microscope's internal reticle. This *zero* micrometer reading is recorded and the object is then moved to align its other extreme. The difference in micrometer readings will indicate the size of the object measured. With this method the accuracy of the reading is tied directly to the accuracy with which the two extreme settings are made. This accuracy can be improved on if the measurement is made at the internal image plane. This method requires a calibrated moving reticle and is limited to measurements no larger than about half the microscope's field of view. The reticle calibration is accomplished by viewing a precise standard scale (referred to as a *stage micrometer*), placed in the object plane, and establishing the exact relationship between the amount of reticle travel and that fixed standard. That relationship can then be applied to measurements of objects whose size is unknown.

Much has been done in recent years to combine the optics of the microscope with the modern, compact, solid-state TV camera. The camera may be provided with a lens that accepts the output from the microscope's eyepiece. Better yet, the eyepiece can be eliminated and the camera, without lens, can be placed at the image plane of the microscope's objective lens. This microscope–TV camera combination opens up a number of new applications and is particularly valuable when remote viewing and simultaneous viewing by a several viewers is desired.

8.5   The Telescope

The telescope and the microscope have much in common, the major difference being that, unlike the microscope, the telescope is used to view objects that are located at some substantial distance from us, when it is not practical to reduce that distance physically. We find that the origins of the prefixes *micro-* (small) and *tele-* (at a distance), are tied to these conditions. Two typical telescope applications would be, viewing the moon from our backyard, and viewing the batter at a major league ball game when our tickets are for the center-field bleachers, some 500 ft from home plate.

By discussing the design of telescopes to perform these two functions we can learn a lot about the optical engineering that would be involved. For observation of the moon (and other astronomical subjects) we might decide on a simple telescope, made up of a refracting objective lens and an eyepiece. A practical limit (based primarily on cost) for the size of an objective lens is about 100 mm diameter, with a speed of $f/10$ (EFL = 1000 mm). To make this design exercise more interesting, let's assume that we have located a source for a family of good, reasonably priced, interchangeable, eyepieces with focal lengths ranging from 8 mm to 28 mm, each designed to cover a 45° field of view.

Our initial design exercise will couple our 1000-mm objective with the 28-mm eyepiece. Telescope magnification is found by dividing the objective EFL by the eyepiece EFL. In this case the result is 1000÷28 = 35.7×. The angular field of view that we will see is limited by the diameter of the field stop in the eyepiece, which manufacturer's data tell us is 23.3 mm. The resulting field of view would be

$$\text{Field of view} = \tan^{-1}\frac{23.3}{1000} = 0.023 = 1.33°$$

The telescope's exit pupil diameter will be equal to its entrance pupil divided by the magnification. For this system that would be 100 mm÷35.7 = 2.8 mm. These characteristics, as outlined in Fig. 8-5, pretty much define the dimensional parameters of the telescope. Let's consider now the optical performance limitations.

For an astronomical telescope it is reasonable to strive for near diffraction limited performance of the objective lens. In this case we can apply the familiar formula to determine the radius of the diffraction limited blur circle (the Airy disk):

$$\text{Airy disk radius } R = 1.22 \times \lambda \times f\text{ number}$$

where λ is wavelength and f number is lens speed. For this objective lens

$$R = 1.22 \times 0.00056 \times 10 = 0.0068 \text{ mm}$$

Design analysis indicates that a well corrected, achromatic, cemented doublet, made from ordinary optical glass types and optimized for

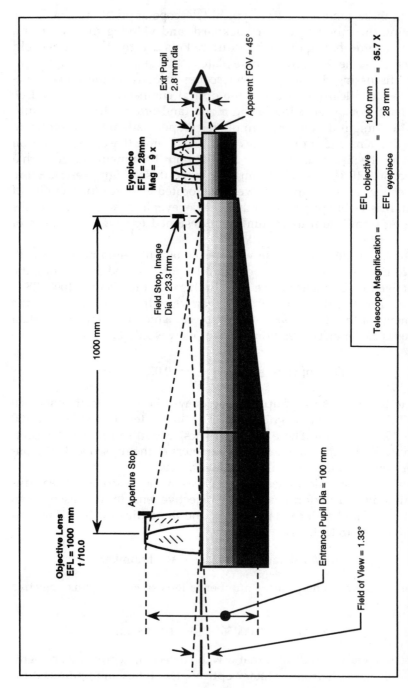

Figure 8-5. Basic optical parameters of an astronomical telescope consisting of a 1000-mm, $f/10$ objective, and a 28-mm eyepiece.

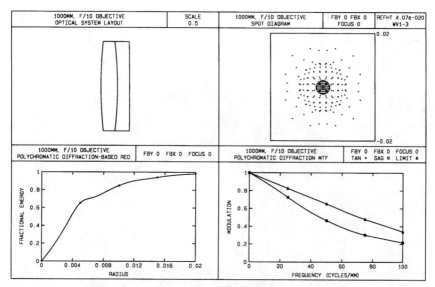

Figure 8-6. Layout and on-axis performance data for a 1000-mm, $f/10$ achromatic cemented doublet.

visual use, will have a spot diagram, radial energy distribution and MTF as shown in Fig. 8-6. The MTF curve seen here is indicative of a lens that will perform essentially to the diffraction limit. The small loss in modulation, relative to the diffraction limit, is due to residual secondary color. Elimination of that aberration can be achieved only by the use of unusual (and expensive) glass types, or through the use of a reflecting design rather than refracting. Neither of these solutions would be reasonable in this instance.

The image formed by the objective will be viewed with a 28-mm (9×) eyepiece. This will enable the average viewer to resolve about 60 cycles/mm at that image. The MTF curve for the objective lens (Fig. 8-6), indicates that the image will have a modulation of about 0.4 at 60 cycles per mm. Any modulation value greater than 0.1 represents a pattern that will be easily resolved by the viewer. Each element of this 60-cycle pattern will subtend an angle of 1.7 arc seconds in object space. This 1.7-second value is the angular resolution limit of the telescope, when it is combined with the performance of the typical visual system. Another approach to determining the resolution limit would be to divide the eye limit (60 arc seconds) by the telescope's overall magnification (35.7×), which would yield the same 1.7 arc second result.

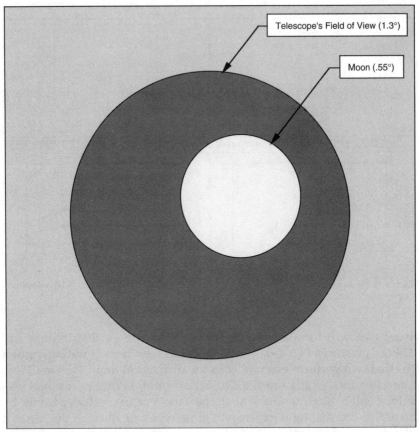

Figure 8-7. The field of view, as seen through a telescope with a 1.3° field of view, looking at the moon.

When the moon is viewed through this telescope, the field of view will be as shown in Fig. 8-7. The moon will subtend an angle to the eye in this case of approximately 20°. Without the telescope, the moon subtends an angle of 0.55°. This angular magnification factor can be described in another way. The distance from the earth to the moon is about 226,000 mi. In order to increase the moons apparent angular size to the naked eye from 0.55 degrees to 20°, we would have to reduce that distance to 6300 mi. In other words, we would have to travel a distance of about 200,000 mi through space toward the moon in order to realize the same view of the moon that has been created using our telescope.

Additional telescope magnification can be realized by introducing the shorter-focal-length eyepieces. The family of commercial eyepieces that had been identified earlier included focal lengths of 28, 21, 15, 12, and 8 mm. When used with the 1000-mm-focal-length objective, these would produce overall magnifications of 35.7×, 47.6×, 66.7×, 83.3×, and 125×, respectively. There are drawbacks to higher magnification. First, any imperfections in the image formed by the objective, due to atmospheric conditions, lens quality, telescope vibration, focus, etc., will be magnified by a shorter focal length eyepiece. Second, higher magnification will generally mean a reduced field of view. In this case, the 1.3° field of view will be reduced to about 0.36° when using the 8-mm eyepiece at 125× magnification. Another characteristic of the telescope design that must be monitored is the diameter and location of the exit pupil. Since we have an $f/10$ objective, it follows that the exit pupil diameter will be $\frac{1}{10}$th the eyepiece focal length. For the 8-mm eyepiece this would mean a 0.8-mm-diameter exit pupil. A pupil this small will begin to adversely affect the resolution capability of the eye. Finally, *eye relief*, which is the distance from the eyepiece to the spot where the eye must be located for optimum viewing, will be roughly equal to the eyepiece focal length. Eye relief that is less than 15 mm is not conducive to comfortable viewing. For all of these reasons, this telescope system should probably be limited to magnifications of 35.7×, 47.6×, and 66.7×. These would be achieved by using respectively the 28-, 21-, and 15-mm-focal-length eyepieces.

Our second telescope application involves viewing a sporting event from a distance of around 500 ft. Why not use our astronomical telescope? There are several very good reasons, its size and weight for starters. That telescope would be some 5 in in diameter and 40 in long and would probably weigh at least 10 lb. Its high magnification would require a tripod mount and, if all that were not enough, the image that we view would be upside down.

For comfortable, handheld viewing, the telescope magnification should not be greater than about 10×. To provide a generous field of view to the eye we might assume an eyepiece with a 50° field of view. For good image brightness we will assume an objective lens (entrance pupil) of 50 mm diameter. Finally, to produce the proper image orientation we will assume that a roof-Pechan prism will be contained in the back focal region of our objective lens. A telescope of this type, designed to view nearby objects, is called a *terrestrial telescope.*

In order to keep the objective lens form reasonably simple it will be given a lens speed of $f/8$. This means the focal lengths of the objective and eyepiece will be 400 and 40 mm, respectively. The objective lens field of view will be 5°, which will result in an internal image size of 35 mm diameter. Knowing that the objective lens diameter is 50 mm and the image diameter is 35 mm, we could reasonably assume that the roof-Pechan prism will require an aperture diameter of about 40 mm. Figure 8-8 shows the resulting 10× telescope system, suitable for viewing of sporting events and similar terrestrial subjects. Telescopes of this type are generally specified in terms of magnification and objective lens size. In this case, this telescope would be designated as 10 × 50, with a field of view of 260 ft at a distance of 1000 yd. Figure 8-9 illustrates the field of view that would be seen through this telescope. The advantage of using this telescope is best appreciated when we consider the fact that the size of the naked eye image of the batter would be just $\frac{1}{10}$th of that shown.

8.6 Binoculars

A pair of binoculars is made up of two identical terrestrial telescopes, linked together such that their optical axes are parallel. The distance between the two exit pupils must be adjustable, to accommodate individual differences in eye separation (interpupillary distance). The standard value for interpupillary distance is 62 mm, with an adjustment range of ±10 mm adequate to satisfy most requirements. It is critical that the two optical axes be parallel as they enter the eyes. A maximum alignment tolerance of about one arc minute will be accommodated by the eyes with little problem. The dual optical paths will result in more natural viewing, with a greatly improved stereo sense, relative to that provided by a single telescope. In general, the improvements offered by binoculars (relative to a similar telescope) will be cosmetic in nature and, in most cases, will not justify the increased size, weight, cost, and complexity involved. The optical layout of a typical pair of 7 × 50 binoculars is shown in Fig. 8-10. The configuration shown is the traditional Porro prism design. Many more recent designs utilize in-line prisms that lead to a more compact, lightweight package.

8.7 The Riflescope

The *riflescope* is a low-power telescope, designed to improve the sighting capability of a rifle. There are several unique characteristics of the riflescope that are not found in other optical instruments. First, it is possible to generate the desired image orientation through the introduction of a relay lens rather than a prism assembly. While this leads to a longer overall length, that length is compatible with the space

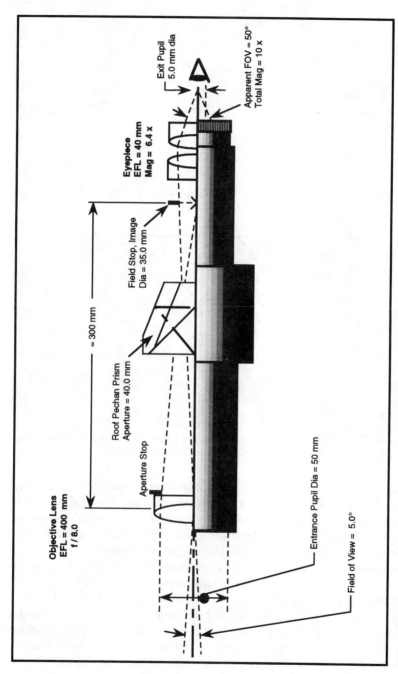

Objective Lens
EFL = 400 mm
f / 8.0

≈ 300 mm

Field of View = 5.0°

Entrance Pupil Dia = 50 mm

Aperture Stop

Roof Pechan Prism
Aperture = 40.0 mm

Field Stop, Image
Dia = 35.0 mm

Eyepiece
EFL = 40 mm
Mag = 6.4 x

Exit Pupil
5.0 mm dia

Apparent FOV ≈ 50°
Total Mag = 10 x

Figure 8-8. Basic optical parameters of a 10 × terrestrial telescope, consisting of a 400-mm, f/8 objective and a 40-mm eyepiece.

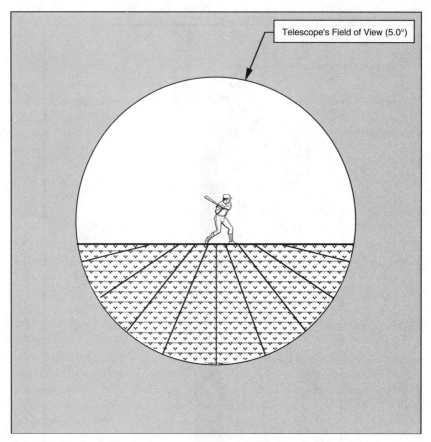

Figure 8-9. The field of view, as seen through a 10 × terrestrial telescope with a 5° field of view, looking at the batter from a distance of 500 ft.

available and actually improves the alignment capability of the rifle-scope. A *reticle,* capable of being adjusted in azimuth and elevation for alignment of the optical axis to the weapon, is an essential component of the riflescope. The optical system must also provide substantial eye relief to prevent trauma to the eye when the rifle is fired.

From the layout in Fig. 8-11 it can be seen that the eyepiece contains the largest optical elements in the riflescope. The overall height of the sight can be reduced if its vertical field of view is reduced. This is accomplished by placing a field stop at the eyepiece image, shaped as shown in the figure. This elongated field of view shape is consistent with the normal (binocular) visual field and it is also compatible with the function of the riflescope. Reduction in the vertical field of view allows the height of the eyepiece elements to be reduced proportion-

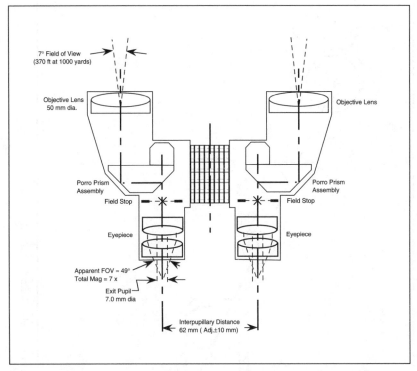

7° Field of View
(370 ft at 1000 yards)

Objective Lens
50 mm dia.

Objective Lens

Porro Prism
Assembly

Porro Prism
Assembly

Field Stop

Field Stop

Eyepiece

Eyepiece

Apparent FOV = 49°
Total Mag = 7 x

Exit Pupil
7.0 mm dia

Interpupillary Distance
62 mm (Adj.±10 mm)

Figure 8-10. Optical layout of a pair of 7 × 50 Porro prism binoculars.

ately. In many modern riflescope designs the relay optics can be adjusted to produce a range of magnifications (zoomed), which adds greatly to the function of the instrument.

8.8 Surveying and Optical Tooling

Many optical instruments are used to make precision measurements of large objects over large distances. Depending on the application, this category of instrumentation has been given the designation of surveying, or optical tooling instruments.

In the field of surveying instruments, the optical sight level and the transit (or theodolite) are the two dominant optical instruments. The optical sight level is a precision, high-power, terrestrial telescope, mounted in such a way that it can sweep a level plane, allowing the user to measure linear departure from that plane. The transit (theodolite) is also a precision, high-power, terrestrial telescope, mounted in

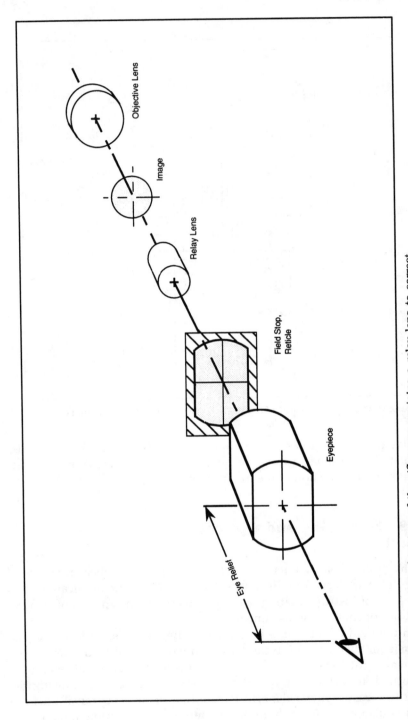

Figure 8-11. The optical system of the riflescope contains a relay lens to correct image orientation and requires larger than usual eye relief.

such a way that its line of sight can be scanned in both azimuth and elevation. It is provided with precision scales to permit the precise measurement of the angles between objects. The major differences between the transit and the theodolite would seem to be size (the theodolite being more compact) and the degree of precision with which angles can be measured (the theodolite generally being more accurate). The optical configuration of the telescope used in both of these instruments will generally contain an objective lens, a focusing lens, a reticle, and an erecting eyepiece assembly that resembles a low-power microscope in its construction and function (see Fig. 8-12).

The field of optical tooling uses surveying instruments as well as a family of special optical tooling instruments that includes collimators, autocollimators, alignment telescopes, and a variety of special-purpose accessories. The collimator contains an illuminated reticle set precisely at the focal plane of the objective lens. The projected (collimated) reticle pattern establishes a reference line of sight to which any number of other optical tooling devices can be aligned, thus establishing their relationship to one another.

The autocollimator is similar to the collimator, but it is equipped with an eyepiece and a beamsplitter for purposes of simultaneous viewing and reticle projection. The process of autocollimation involves projecting the reticle such that it reflects from a flat reference mirror. That reflected reticle pattern is then viewed through the eyepiece of the autocollimator. The reflected pattern will be in precise alignment with the reticle at the eyepiece image plane only when the mirror surface is precisely perpendicular to the line of sight. Another way of describing the process of autocollimation...an image of the autocollimator is formed behind the reference mirror. In order to site squarely into that image, the mirror must be set precisely at a right angle to the optical axis of the autocollimator. Figure 8-13 illustrates the basic collimator and autocollimator.

The alignment telescope [also called *jig alignment telescope* (JAT)] is an autocollimator with the additional capability of making precise linear measurements in the *X-Y* plane, at relatively great distances. This is accomplished by including the capability to focus on objects at all distances, along with an optical micrometer, which permits precise, calibrated displacement of the JAT's projected line of sight. In its most basic form the optical micrometer is a plano parallel window, placed ahead of the objective lens and capable of being tilted through a calibrated angle, thus displacing the line of sight. A micrometer drum is attached to the plate, giving a direct readout of the amount of dis-

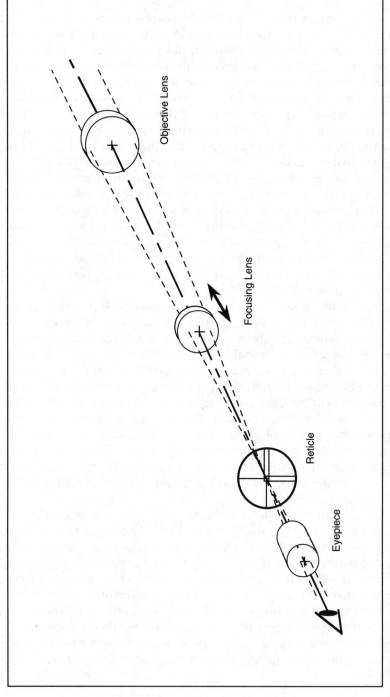

Figure 8-12. The optical layout of a typical telescope system for use in a sight level or transit, two of the primary optical instruments used for surveying.

Objective Lens

Focusing Lens

Reticle

Eyepiece

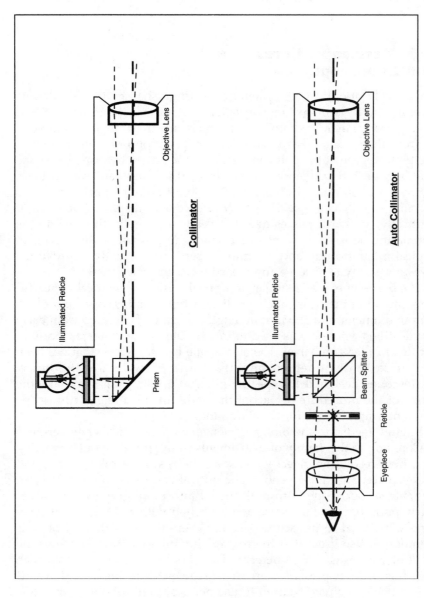

Figure 8-13. Two basic optical instruments used in optical tooling applications.

placement. Figure 8-14 illustrates the optical micrometer and includes formulae for line-of-sight displacement as a function of tilt angle.

8.9 Periscope, Borescope, and Endoscope

This family of optical instruments is distinguished by the fact that its overall length is many times greater (often more than one hundred times greater) than its diameter. The periscope may range in complexity from the cardboard box with flat mirrors at opposite ends, such as we typically see at the golf course, to the modern submarine periscope that may be 7 in in diameter and over 40 ft in length and may contain more than 100 optical elements. The "golf course" periscope does nothing to enhance visual capability, except to displace line-of-sight vertically, so that it passes over the heads of those in the crowd. The submarine periscope, on the other hand, can provide several levels of optical magnification and a family of sensors to greatly improve the viewer's ability to observe and record the scene being viewed.

The borescope, while similar in optical design to the periscope, differs greatly in that it is much smaller. A typical borescope might be 8 mm in diameter and 500 mm in length. Its name is derived from early applications where it was used to inspect the inside diameter (bore) of a rifle or other gun barrel. The endoscope is a borescope that has been specifically designed to penetrate and view inside the body. Endoscopes designed for specific medical applications may take the name of the procedure or part of the body involved (e.g., arthroscope for joints, laparoscope for the abdominal cavity).

Enough medicine and biology; let's get back to optical engineering.

Common to this family of instruments is an optical layout consisting of a high-resolution objective lens, a reticle, several relay lenses, and an eyepiece. Figure 8-15 shows the optical configuration of a modern submarine periscope with both visual and photographic capability. Early periscope designs were accomplished with a single set of relay lenses. That led to periscope designs where the vignetting of off-axis bundles was so great that the relative illumination at the edge of the field of view was just 10 percent. The versatility of the eye was such that this did not present a problem. However, during World War II (early 1940s) the navy started to use periscopes for photographic missions and the poor relative illumination led to poor photographs. The solution was to add two more sets of relays (one additional set would have produced an inverted image) to the optical design. In this config-

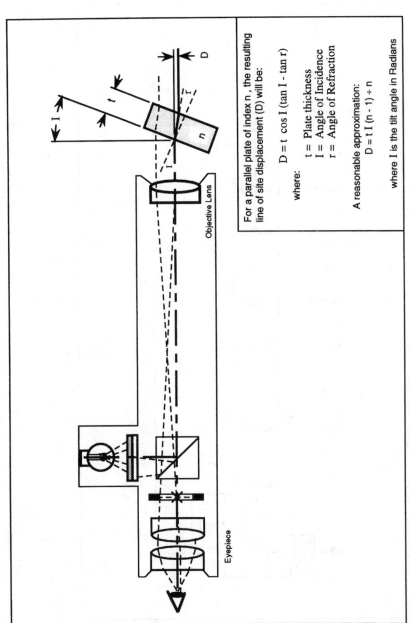

For a parallel plate of index n, the resulting line of site displacement (D) will be:

$$D = t \cos I (\tan I - \tan r)$$

where:
- t = Plate thickness
- I = Angle of Incidence
- r = Angle of Refraction

A reasonable approximation:
$$D = t I (n - 1) \div n$$

where I is the tilt angle in Radians

Objective Lens

Eyepiece

Figure 8-14. The optical micrometer is a plano parallel plate which, when tilted as shown, will displace the line of sight by an amount *D*.

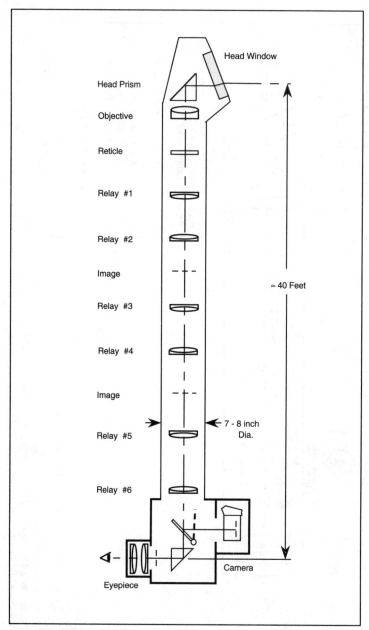

Figure 8-15. Optical layout of a modern submarine periscope system. Elevation scan is accomplished by the head prism, while the viewer rotates the entire periscope about its vertical axis to achieve azimuth scan.

uration the relative illumination was increased to 40 percent, a variation that could be handled by the exposure tolerance (latitude) of most films. Other popular modes commonly incorporated into the submarine periscope include TV and low-light-level imaging. Early periscopes often included an optical means of estimating range to the target. This usually required knowledge of the target size and a means of measuring image size. In recent years this ranging process has become simpler and more accurate with the incorporation of the laser range finder into the periscope's optical system.

The borescope is quite similar to the submarine periscope, with the major exception of its size. While the diameter of the borescope may typically range from 4 to 12 mm, the lenses that make up the borescope optical system may be as small as 1 mm in diameter. Figure 8-16 shows a typical optics layout for a borescope. Along with the viewing optics, the borescope shown also contains a coaxial bundle of fiber optics to deliver illumination to the field being viewed. It will be noted in the figure that the lenses in the borescope do not conform to the usual proportion of length to diameter (typically 1:8). Because these optics are so small in diameter, the relative thickness must be increased substantially to facilitate manufacture and to maintain alignment within the final assembly. While the 90° head prism shown is a popular configuration, it is possible to eliminate the head prism to permit straight ahead viewing. Other prism configurations make it possible to view at other angles.

The rapidly developing field of compact, solid-state electronics for imaging has caused several major changes in the field of periscopes, borescopes, etc. The developing trend for all of these instruments is to place the detector as close as possible to the head prism. This often eliminates the need for relay optics and will, in many cases, result in improved image quality. Once the image has been formed and converted to an electronic signal, it can be transported via wire, or optical fiber, to a remote location for viewing, broadcast or recording. In the submarine periscope version of this configuration, a compact optical system is used in conjunction with high-resolution monochrome and color TV cameras in the head of the periscope. Camera output is transmitted to the ship's command center, where it is displayed on a bank of high-resolution monitors.

In the case of the borescope for medical applications, this approach has been accomplished by what is called the "chip in the tip" approach. Here the image is formed, using a high precision micro objective lens, on a solid-state chip that may be as small as 1.5 mm square. Output from the chip is transmitted to auxiliary electronics outside of the borescope where it is converted to a conventional TV

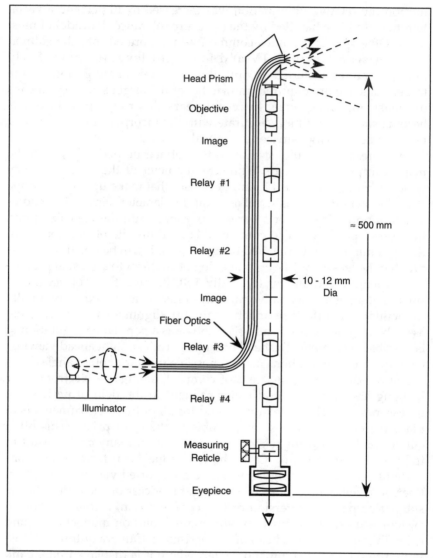

Figure 8-16. Optical layout of a typical borescope system. Unique to the borescope is the requirement that it contain an integral light source to illuminate the object being viewed.

signal for viewing by the medical team. Several techniques have been developed recently that permit the forming of dual images which, when viewed under the proper circumstances, present a true three-dimensional, stereo image to the observer. The advantage of this approach to certain medical procedures has been significant.

In certain applications it is advantageous to have a flexible borescope rather than a rigid assembly. This has been accomplished using a flexible coherent fiber-optics bundle to transfer the image from the tip of the instrument to the output end.This approach will be limited by the resolution of the fiber bundle and its light transmission efficiency. More recently, the electronic approach has made it possible to form the image in the tip and then transfer it electronically, through a flexible cable, to an external point for further processing.

While many of these recent developments have led to optically simpler systems, with reduced numbers of lenses, the compact nature and precision requirements of the resulting optical systems have challenged, and will continue to challenge, the skills and expertise of the optical engineer and lens designer.

8.10   Review and Summary

This chapter has presented a review of the field of basic optical instruments. The primary intent has been to familiarize the reader with some of the optical engineering considerations that are involved in the design of such instruments. From the simplest magnifier to the rather complex submarine periscope, it is the optical designer's responsibility to understand the intended function of the instrument and to ensure that the optical system is capable of carrying out that function. Most important is the final evaluation of the instrument's image quality. In many cases the eye is the system detector and as such must be included in any performance evaluation. Several examples are given here demonstrating how that might be done. In a subsequent chapter more detailed information on the characteristics and performance of the human visual system will be presented.

In several instances this chapter has indicated the impact that modern technological developments, particularly in the area of electronics and detectors, have had on instrument design. The laser range finder and compact solid-state TV camera are two examples of components that have produced significant changes in recent years. The lesson offered here is that it is extremely important that the optical engineer remain current on technological developments and be prepared to explore methods of including them in instrument design.

9

Optical Materials and Coatings

9.1 Introduction

This chapter will be devoted primarily to discussing a variety of materials that are used in the design and manufacture of optical components. The most common of these would be optical glass, in its many forms. Other materials, such as plastic and fused silica, are sometimes substituted for optical glass in order to take advantage of their special characteristics. Optical coatings are added to components to enhance their optical performance. Antireflection coatings will improve the light transmission of a lens, single- and multiple-layer antireflection coatings will be covered. Materials for use in infrared systems are unusual in several respects. The basic characteristics of the most commonly used IR materials will also be presented.

All these topics will be covered in an introductory fashion, with the intent to provide you with a level of knowledge that will permit you to generate basic specifications and discuss requirements with specialists in these fields.

9.2 Optical Glass

The material we refer to as *optical glass* differs from the more common glasses in that its make up and ultimate performance characteristics are precisely controlled during all stages of the manufacturing process. This permits the optical designer to generate a design configuration based on published catalog data that, when manufactured, will perform exactly as indicated by the design data.

There are several reliable, established sources for optical glass throughout the world. Primary sources manufacture a complete variety of precision optical glasses from basic raw materials. Other glass suppliers act as processors, taking bulk supplies of glass from the original manufacturers and modifying the size and shape of starting glass blanks, while precisely controlling the characteristics of the glass during this process. It is enthusiastically recommended that the interested reader contact all glass manufacturers and suppliers, requesting catalog and technical data describing their products and services. With this information available, it will then be a simple matter to select the most appropriate source to meet your requirements.

Much of the information that follows will refer to data and materials supplied by the Schott Optical Glass Company of Duryea, Pennsylvania (USA) and Schott Glaswerk of Mainz, Germany. I have opted to present this information because Schott is the glass company with which I have had the most personal experience and, as a result, have collected the most data and general information. This choice on my part is not intended to indicate a superiority of Schott over the many other fine, established suppliers of optical glass.

Optical glass must have several unique characteristics in order to perform its required function. Optical glass must be transparent to the wavelengths of the design and its index of refraction over the spectral band of the design must be known with a high degree of precision. Beyond these critical factors, it is important to take into account such factors as the cost and *workability* of the glass, along with its many physical, mechanical and thermal characteristics.

Individual optical glass types are identified by both name and number. The name is assigned by the manufacturer and will generally identify the glass type based on the key elements that it contains. For example, typical Schott glass names are BK (borosilicate crown) and LaF (lanthanum flint). Other glass producers have developed their own methods for naming glasses. As a result, the same glass type may be known by several different names, depending on the source.

The glass identification number is a six-digit code derived from the index of refraction and dispersion of the glass. If, for example, the index of a glass is 1.517 and the dispersion is represented by an Abbe v number of 64.2, the numerical identifying code for that glass would be 517642. The glass identification number has the obvious advantage that it is not optional on the part of the glass manufacturer, but is tied directly to the optical characteristics of the glass, regardless of the source.

The glassmap shown in Fig. 9-1 indicates the index and v number for 65 of today's commonly used optical glass types. A comprehensive glass map, containing these, along with several hundred additional glass types, can be obtained from Schott or any of the other primary glass producers. The *select* glass types shown in Fig. 9-1 have been selected for their ready availability, reasonable cost, and good workability characteristics. A logical lens design procedure might start using these *select* glasses and then depart into other more exotic glass types, only when required. Any departures from the select glasses should be carefully monitored to ensure that the design remains reasonable in terms of cost and producibility.

The Abbe v number has been referenced as an indicator of the dispersion characteristic of the glass. The v number is calculated using the following formula:

$$v_d = \frac{n_d - 1.0}{n_f - n_c}$$

where n_d, n_f, and n_c are the indices at 0.588 μm (yellow), 0.486 μm (blue), and 0.656 μm (red), respectively.

This v_d number for an optical glass indicates the amount of dispersion that will occur across the visible portion of the spectrum. As Fig. 9-1 shows, the v number for the select glasses will range from 25 to 70, with common values for crown glass being about 60 and flint glass about 30. It will also be seen from the glassmap that the index of refraction for these most frequently used glass types will range from 1.50 to 1.80.

The significance of index of refraction change with wavelength (dispersion) is shown in Fig. 9-2. For this example, data for two simple lenses, one made from common crown glass (BK7) and the other from a common flint glass (SF1), are shown. Both lenses have a nominal focal length of 100 at a wavelength of 0.5876 μm. The curves show how the focal lengths of these lenses will change when they are used at other wavelengths. Over the wavelength range from 0.35 to 1.1 μm, the focal length of the crown lens will vary from −4 to +2 percent, while the flint element focal length changes from −10 to +4 percent.

In order to generate a proper and functional optical design, it is essential that many other characteristics (beyond index and dispersion) of the glasses used be carefully considered. The light transmission of most optical glasses will be quite high for wavelengths from 0.4 to 2.0 μm (see Fig. 9-3). When system transmission is critical near these extremes of wavelength, more detailed information on the particular glass types being considered should be referenced. There is a wealth of additional information available from the glass manufacturers detail-

MIL Type	Schott Type		MIL Type	Schott Type
501564	K10		626390	BASF1
510635	BK1		636353	F6
511604	K7		639554	SK18A
515547	KF3		641601	LAK21
517642	BK7		648339	SF2
521697	PK50		650392	BASF10
522595	K5		651559	LAKN22
523515	KF9		652449	BAF51
532488	LLF6		652585	LAKN7
540597	BAK2		658509	SSKN5
547536	BALF5		664358	BASF2
548458	LLF1		670471	BAFN10
552635	PSK3		673322	SF5
564608	SK11		678552	LAKN12
569561	BAK4		683445	BAF50
573575	BAK1		689312	SF8
575415	LF7		689497	LAFN23
580537	BALF4		699301	SF15
581409	LF5		702410	BASF52
583465	BAF3		713538	LAK8
589613	SK5		717295	SF1
603380	F5		717480	LAF3
603606	SK14		720504	LAK10
606439	BAF4		724381	BASF51
607567	SK2		728287	SF53
609464	BAF52		744447	LAF2
618498	SSKN8		750350	LAFN7
620364	F2		755276	SF4
620603	SK16		762265	SF14
622532	SSK2		782372	LAF22A
623569	SK10		785261	SF56A
623581	SK15			
624470	BAF8			
626357	F1			

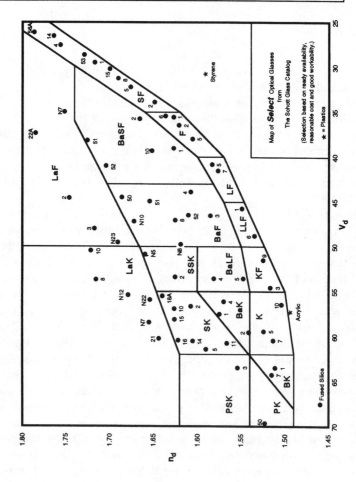

Figure 9-1. Glassmap and list of *select* optical glasses.

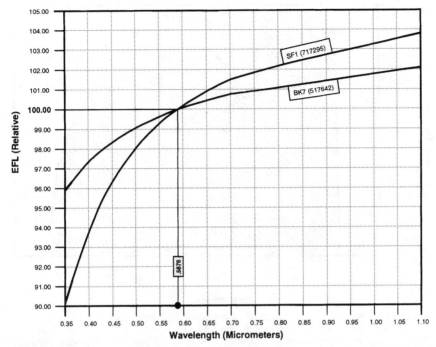

Figure 9-2. Because the index of refraction is not constant, the focal length (EFL) of a simple lens will vary as a function of wavelength. Shown here are representative curves for lenses made from typical crown and flint optical glasses.

ing the optical, mechanical, and thermal characteristics of its glasses. The optical designer should carefully review all this information and consider it with respect to the detailed requirements of the particular system being designed.

Catalog data will indicate the nominal index of refraction for each glass at a number of wavelengths. Normal procurement will result in glass that is within ±.001 of these catalog index values, with a ±0.8 percent tolerance on the v number. The designer must determine if this level of index control is consistent with system requirements. When required, the index can be controlled to as close as ±0.0002, with a corresponding v number tolerance to ±0.2 percent. These reduced tolerances will result in a slight increase in the cost of the glass. When ordering glass it is usually possible to obtain melt data from the glass supplier, containing precise measured index of refraction data for the particular glass blanks that have been delivered.

In addition to the basic index value for any glass blank, it is equally important to control the homogeneity throughout that blank.

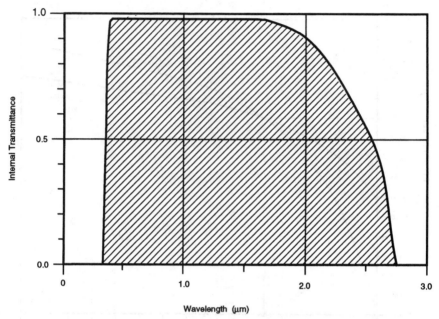

Figure 9-3. Internal transmittance of a typical optical glass with a thickness of 25 mm. The shape of the curve at cuton and cutoff will vary for several unusual glasses. The value for internal transmittance at 0.4 μm will be found on data sheets for most glasses, a low value indicating a yellowish color for thick glass blanks.

Homogeneity is the amount of variation in index of refraction that is allowed throughout the volume of any one glass blank. The standard (default) value for homogeneity within a melt is $1 \cdot 10^{-4}$. The homogeneity of any individual piece taken from that melt will be less. When required, it is possible to produce glass blanks with homogeneity values as small as $5 \cdot 10^{-7}$. Homogeneity becomes particularly important in optical components that are part of a near-diffraction-limited system.

Bubbles, seeds, and *striae* are categories of internal glass defects that may be observed, usually under very sensitive viewing conditions. Bubbles are formed when small air pockets are left as the glass cools. Seeds and striae are solid contaminants within the glass; seeds are near-spherical, while striae are long and stringlike. Except for a few very unusual cases, most of today's optical glasses will not contain bubbles, seeds, or striae to any significant degree. Optics designed to transmit high-power lasers may require special consideration regarding bubble and seed content.

The thermal characteristics of the glass must be considered when the temperature range over which the lens is to be used is substantial. There are two values that enter into thermal considerations. They are the coefficient of linear expansion α and the change in index with temperature dn/dt. The linear expansion of an optical element will most generally be acceptable if that element has not been mechanically constrained such that severe stress or possible breakage of the element might occur as a result of temperature changes. The change in index with temperature dn/dt will produce a focus shift and possible significant degradation of image quality if refocus is not possible. This change in focus can be analyzed and, in some cases, compensated for by the proper choice of lens cell material. This area of system design is a joint responsibility, to be shared by the optical and the mechanical design engineers.

9.3 Low-Expansion Materials

In a number of applications it is desirable to use optical materials that have a low coefficient of thermal expansion. Most common of these is the case of the first-surface mirror blank. One economical choice in this application is the material produced by Corning, with the trade name *Pyrex*. This is a very stable optical material with a thermal expansion that is less than half that of typical optical glasses. Pyrex is not suitable for most lens applications due to internal coloration and defects, such as bubbles and seeds.

Lenses and higher-quality mirrors can be made from *fused quartz* (also called *fused silica*). Fused quartz is optically pure and has a coefficient of expansion that is less than $\frac{1}{10}$th that of typical glass. When even greater thermal stability is desired there are several materials with expansion coefficients that are essentially zero (Schott Zerodur, Corning ULE, and Owens CERVIT). As with Pyrex, these materials are not suitable for use as lens blanks.

It is interesting to note that, while fused quartz is an ideal lens blank material in terms of its optical quality and low expansion coefficient, it does have a relatively high dn/dt characteristic. The result of this is that, in a lens application, while the size and shape of the lens will remain essentially constant with change in temperature, the focal length of the lens will fluctuate with the change in index of refraction. The curve shown in Fig. 9-4 illustrates the change in focal length with temperature for a lens element made from fused quartz as compared with a similar lens made from BK7 optical glass. In the case of the fused-quartz lens the dn/dt is essentially the sole cause of the focal length change while for the BK7 element the linear expansion of the lens (with increased temperature), which increases its

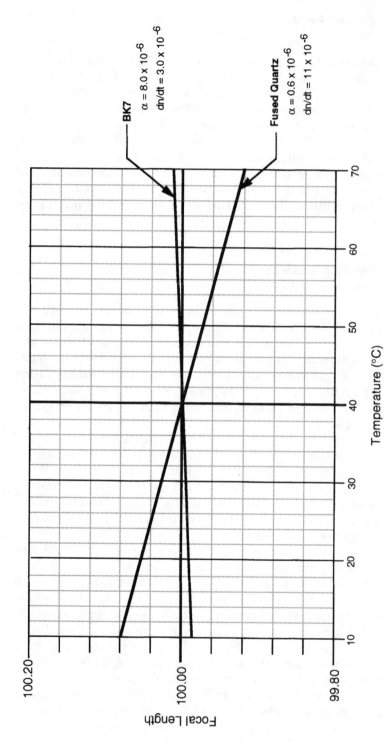

Figure 9-4. The focal length of an optical element will vary as the ambient temperature changes. Two factors are involved: the linear expansion α of the lens material and the change in index of refraction with temperature dn/dt.

EFL, is partially offset by an increase in the index of refraction, resulting in a net focal length change that is considerably less than that of the fused-quartz lens. This is a very basic demonstration of how a change in temperature can effect the focal length of a lens element. Any optical system that is subject to substantial temperature fluctuations must be carefully analyzed, taking into account the expansion coefficients of all optical and mechanical parts, combined with the dn/dt of the refracting optical materials. Many of today's computer programs for optical design include routines written to perform this type of analysis.

9.4 Surface Losses and Antireflection Coatings

When light passes into or out of an optical component there will be a reflection loss at the air–glass interface. The amount of loss will be proportional to the index of refraction of the glass from which the component is made. The general formula used to compute the surface reflection loss (air to glass) is as follows:

$$R = \frac{(n - 1)^2}{(n + 1)^2}$$

For a standard optical glass with an index of refraction of 1.5, this formula indicates a loss of about 4 percent. For glass with an index of 1.8, the surface reflection loss will reach 8 percent. When several surfaces are involved it is important to remember that the absolute loss per surface becomes slightly less at each surface. Assume that we have five elements made from glass with an index of 1.8. The final transmission, ignoring absorbtion losses, through the 10 air to glass surfaces would be found using the following formula:

$$t = 0.92^{10} = 0.43 \quad \text{or} \quad 43\%$$

As lens designs have became more complex, involving the use of more elements and high-index glasses, the need to reduce surface losses has become critical. It was discovered that if a very thin layer of the properly chosen transparent material were added to the glass surface, those losses could be reduced substantially. The key to a successful antireflection (AR) coating was found to be selecting a material with an index of refraction that was equal to the square root of the glass

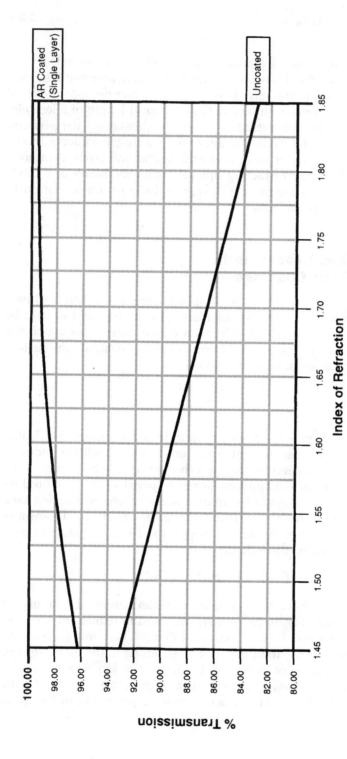

Figure 9-5. Transmission curves (at $\lambda = 0.588$ μm) through a window or lens, as a function of index of refraction. The lower curve is for an uncoated element, while the upper curve is for an element with a single-layer AR coating of magnesium fluoride on both sides.

index, and controlling the thickness of the layer to be exactly $\frac{1}{4}$ of the wavelength being transmitted. Figure 9-5 shows the transmission of a single element, uncoated and coated with a single layer of AR coating, as a function of the element's index of refraction.

It develops that rarely can either or both of these criteria be met exactly. The selection of available coating materials is quite limited, making the index of refraction match quite difficult. The $\frac{1}{4}$ wave thickness is not a problem for a monochromatic system. However, for the more common system, designed to cover a broad wavelength range, a compromise thickness must be selected, usually based on the center of that range. The development of multilayer coatings has made it possible to accomplish most reasonable transmission goals with just a slight addition of cost and complexity to the final lens assembly. The data shown in Fig. 9-6 illustrates the effectiveness of single- and multilayer AR coatings on the transmission of typical lens elements with both surfaces coated. It will be noted that for glass of normal index (about 1.5), the improvement in transmission is about 6 percent when a single-layer coating is used, with an additional 2 percent gain possible when the multilayer coating is used. For the case of higher-index glass (about 1.7), the single-layer coating results in a gain of about 10 percent, while the multilayer coating permits an additional 1 to 2 percent gain. In general, for systems operating over the visible spectrum, single-layer coatings are most cost effective on high-index glasses, while for glasses in the normal-index range, multilayer AR coatings are usually the best choice.

9.5 Materials for Infrared Systems

In recent years many applications have been found for optical systems designed to operate in the infrared portion of the spectrum. Such systems are designed to sense thermal radiation emitted by the object being viewed. As such, it is possible to view objects that are not illuminated, and may not be visible, in the traditional sense. The most common applications of IR systems involve night vision. Most of the useful IR energy emitted by warm or hot objects will have wavelengths in the 3- to 12-μm spectral range. For systems that are operating in the earth's atmosphere, much of this energy will not be transmitted. As Fig. 9-7 shows, the atmosphere transmits two distinct IR bands, one from 3 to 5 μm and the second from 8 to 12 μm. This section will deal with four of the most popular materials used in IR optical system

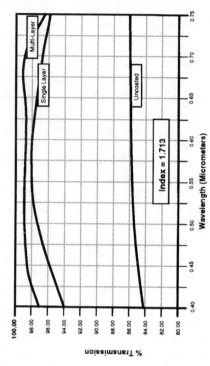

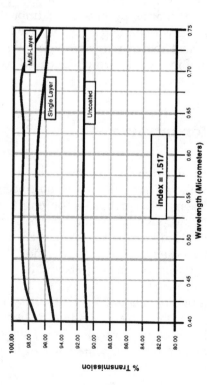

Figure 9-6. Transmission curves through a 5-mm-thick window, made from normal index optical glass (517642) and high-index glass (713538), uncoated and AR-coated with a single layer and a multilayer coating.

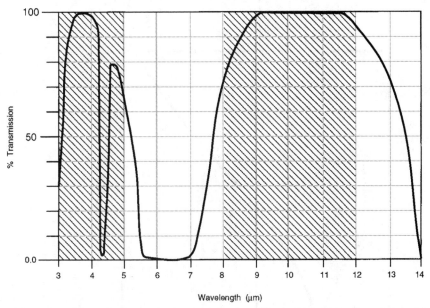

Figure 9-7. Smoothed curve, showing the two IR windows (3 to 5 and 8 to 12 µm), created by the average transmission characteristics of the earth's atmosphere (approximate path length 0.5 km, at sea level).

designs covering these spectral bands. These are germanium, silicon, zinc sulfide, and zinc selenide.

Germanium

Germanium (Ge) is a popular material for lenses designed to transmit IR energy in either the 3 to 5 or 8 to 12-µm spectral bands. Germanium has an index of refraction of approximately 4.0 at these wavelengths. Because of this high index, the surface reflection losses for an uncoated air–germanium surface will be so great (≈36 percent) as to make the use of an uncoated germanium optic totally impractical. A germanium window that is uncoated will transmit only 47 percent, including all internal reflections. Broadband AR coatings for germanium will improve the transmission of a single element to around 93 percent over the 3 to 12-µm range. Coatings designed for either the 3 to 5 or the 8 to 12-µm spectral bands will increase element transmission to about 98 percent.

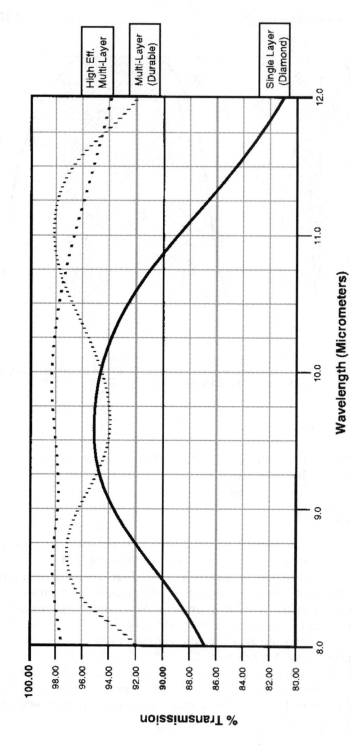

Figure 9-8. Transmission curves through a germanium window with a single layer AR coat (diamond), durable multilayer and high-efficiency (less durable) multilayer AR coating on both sides.

There are three popular AR coating designs used on germanium optics, depending on the environment to which the component will be exposed. Figure 9-8 shows the transmission curves for a single germanium element coated with a very durable single layer (diamond coat), a durable multilayer coating and a less durable multilayer coating suitable for components that are protected from harsh environmental conditions and not subjected to frequent cleaning.

Germanium has mechanical characteristics and appearance that are more like metal than glass. With a density of 5.32 g/cm^3, germanium weighs about twice as much as glass. Its metallic characteristics make germanium a good candidate for the micromachining process. This makes the production of aspheric surfaces on germanium lenses a relatively straightforward matter. This feature, combined with its high index of refraction, leads to many unusually simple but effective optical designs for IR systems using germanium.

Silicon

Silicon (Si) is a popular material for lenses designed to transmit IR energy in the 3 to 5-μm spectral band. Silicon has an index of refraction of approximately 3.43 at these wavelengths. Because of this high index, the surface reflection losses for an uncoated air–silicon surface will be so great ($\approx$30 percent) as to make the use of an uncoated silicon optic totally impractical. A silicon window that is uncoated will transmit about 54 percent, including all internal reflections. A single-layer AR coating for silicon will improve the transmission of a single element to around 90 percent in the 3 to 5-μm spectral band. With a density of 2.33 g/cm^3, silicon weighs about the same as glass. Silicon lenses can be produced using the micromachining process, making the production of aspheric surfaces a relatively simple matter.

Zinc Sulfide

Zinc sulfide (ZnS) is a popular material for lenses designed to transmit IR energy in either the 3 to 5 or 8 to 12-μm spectral bands. Zinc sulfide has an index of refraction of approximately 2.2 at these wavelengths. Because of this high index, the surface reflection losses for an uncoated air to zinc sulfide surface will be so great ($\approx$14 percent) as to make the use of an uncoated zinc sulfide element impractical. A zinc sulfide window that is uncoated will transmit about 75 percent, including all multiple reflections. A broadband AR coating for zinc sulfide will improve the transmission of a single element to around 93 percent over the entire 3 to 12-μm band. Coatings designed specifically for either the 3 to 5 or the 8 to 12-μm spectral band will improve single-element transmission to about 98 percent. Zinc sulfide has mechanical characteristics and appearance that are similar to optical glass. With a density of 4.1 g/cm^3, zinc sulfide weighs nearly twice as much as the average optical glass. The internal transmission of normal zinc sulfide is good over the spectral band from 3 to 12 μm. In addition, it is possible to treat normal zinc sulfide such that its transmission window is increased to include the visible portion of the spectrum. The resulting material is known as *multispectral-grade zinc sulfide* and also by the trade name of *Cleartran* (see Fig. 9-8). This improved visible transmission is very helpful in the alignment of lenses made from zinc sulfide.

Zinc Selenide

Zinc selenide (ZnSe) is another popular material for lenses designed to transmit IR energy in either the 3 to 5 or 8 to 12-μm spectral bands. Zinc selenide has an index of refraction of approximately 2.4 at these wavelengths. Because of this high index, the surface reflection losses for an uncoated air to zinc sulfide surface will be so great (≈17 percent) as to make the use of an uncoated zinc sulfide element impractical. A zinc selenide window that is uncoated will transmit about 71 percent, including all internal reflections. Antireflection coatings for zinc sulfide will improve the transmission of a single element to around 93 percent over the broadband from 3 to 12 μm. Other coating designs will bring the transmission to 98 percent in either the 3 to 5 or the 8 to 12-μm spectral bands. Zinc selenide has mechanical characteristics and appearance that are similar to optical glass. With a density of 5.27 g/cm^3, zinc sulfide weighs nearly twice as much as the average optical glass. The transmission of zinc selenide covers the spectral band from 0.6 to 20 μm. While appearing slightly yellowish in color, zinc selenide does transmit enough visible energy to permit relatively simple alignment of optical elements. Zinc selenide is noted for its superior optical properties, which include very low absorption at the infrared wavelengths.

Figure 9-9 contains the basic transmission curves for all of these IR materials. Note that these curves are for uncoated elements, indicating the substantial surface reflection losses that will occur.

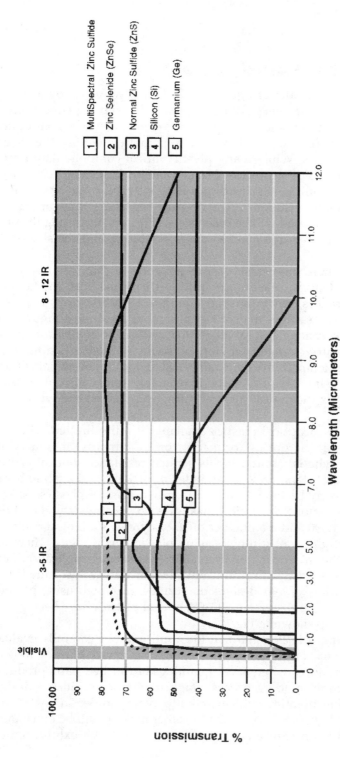

Figure 9-9. Transmission curves for the four most commonly used IR materials. Special treatment of zinc sulfide results in increased visible transmission for the multispectral-grade (also known as *Cleartran*). In regions of maximum transmission, losses are due primarily to surface reflections.

1   MultiSpectral Zinc Sulfide

2   Zinc Selenide (ZnSe)

3   Normal Zinc Sulfide (ZnS)

4   Silicon (Si)

5   Germanium (Ge)

9.6 Optical Plastics

Optical-quality plastics bring several distinct advantages to the design process. Plastics are very strong, capable of surviving conditions of severe shock that would destroy similar elements made from optical glass. Plastics are also light in weight, nominally one third the weight of normal glasses. While many plastic optical components are produced by the molding process, plastic can also be worked using conventional machining steps with normal tools, followed by modified optical polishing techniques. The decision regarding the manufacturing process to be used will be based several factors, including the optical quality required, the quantity to be produced, and the size of the components.

As usual, there does exist a set of tradeoffs to counter these many advantages. Optical plastics are quite difficult to finish to standard optical tolerances. Once finished, the optical stability of the plastic component is less than that of a comparable glass part. Figure 9-10 illustrates the change in focal length that will occur in a plastic lens as the temperature changes. Obviously, some method of maintaining focus during temperature shift (athermalization) must be part of any lens design using plastics. Plastic is relatively soft, which often leads to scratching of the optical surfaces during production and when parts are cleaned.

The variety of optical plastics available is quite limited. The vast majority of plastic optics are made from polymethyl methacrylate (acrylic). Acrylic is favored for its scratch resistance, optical clarity, and mechanical stability. With an index n_d of 1.491 and a v number of 57.2, acrylic may be thought of as the crown glass of plastics. Second in popularity among the optical plastics is polystyrene (styrene). With an index of 1.590 and a v number of 30.8, styrene is a logical choice to act as a flint glass substitute in many plastic lens designs. There are just a few other plastic materials available for special applications, but in most cases acrylic and styrene will be the materials best suited. This limits the variety of lens designs that can be generated using plastics. Acrylic (491572) and styrene (590308) optical plastics have been added to the glassmap shown in Fig. 9-1.

In summary, it is essential that the application be carefully evaluated when the use of plastic optics is being considered. Plastics are well suited for instances where large quantities of finished parts produced by the injection molding process might be involved. Such a design, incorporating plastics, will nearly always be lower in cost, more rugged, and lighter in weight. On the other hand, it will be much more sensitive to temperature changes and will generally exhibit optical

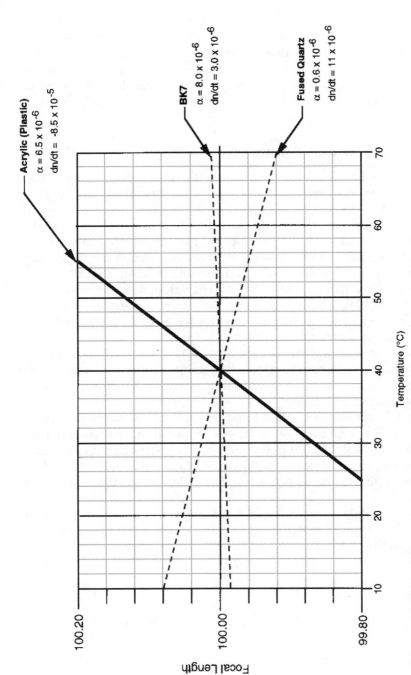

Figure 9-10. The focal length of an optical element will vary as the ambient temperature changes. In the case of acrylic (plastic), the rate of change is affected by large thermal constants (α and dn/dt) that add rather than cancel.

quality that is considerably less than that of a similar glass design. Quite often a very successful design will result when the beneficial characteristics of both optical plastic and optical glass are combined. Such a design might incorporate glass elements in areas exposed to weather and frequent cleaning, while plastics can be introduced to reduce weight and introduce low-cost, molded aspheric surfaces.

9.7 Review and Summary

This chapter has presented the reader with an introductory discussion of the special materials that are used in the manufacture of optical components. The section on optical glass has identified a *select* set of glasses, that are particularly well suited to the majority of lens designs. While several hundred different optical glass types are available, the select list consists of 65 glasses that have been chosen for their ready availability, good workability, and reasonable cost. The section on low-expansion materials describes the characteristics of several optical glasses that have a very low coefficient of thermal expansion. While some of these materials may be used for lenses and prisms, their primary use is as a substrate for first-surface mirrors.

Whenever light passes through an air-glass interface there will be losses due to surface reflections. The subject of antireflection (AR) coatings has been presented, with data detailing the effectiveness of these coatings.

For infrared systems, special and unique optical materials will be required. The most common of these IR materials have been described and their most significant optical and physical characteristics presented. Finally, the subject of optical-grade plastics has been covered. A discussion of the pros and cons of plastics in optical system design is included.

The table of data shown in Fig. 9-11 summarizes the most significant properties of the IR materials and optical plastics that have been discussed in this chapter.

Material	Symbol	Density (g/cm³)	Thermal Exp. (α/°C)	dn/dt (/°C)	Index of Refraction (Wavelength in μm)										
					.586	.485	.656	3.0	4.0	5.0	8.0	10.0	12.0
(IR)													
Germanium	Ge	5.32	6.0 x 10⁻⁶	40.8 x 10⁻⁵				4.045	4.0245	4.016	4.0055	4.0032	4.0020
Silicon	Si	2.33	4.2 x 10⁻⁶	3.9 x 10⁻⁵				3.432	3.425	3.422	3.4184	3.4179	3.4176
Zinc Sulfide (Multi Spectral)	ZnS	4.09	7.9 x 10⁻⁶	4.1 x 10⁻⁵	2.371	2.439	2.342	2.257	2.252	2.246	2.223	2.200	2.170
Zinc Selenide	ZnSe	5.27	7.6 x 10⁻⁶	6.1 x 10⁻⁵	2.631		2.576	2.438	2.433	2.430	2.417	2.407	2.393
(Plastics)													
Methyl Methacrylate	Acrylic	1.19	6.5 x 10⁻⁵	-8.5 x 10⁻⁵	1.491	1.497	1.489						
Polystyrene	Styrene	1.06	6.3 x 10⁻⁵	-12.0 x 10⁻⁵	1.590	1.604	1.584						

Figure 9-11. Table of significant optical and mechanical properties of those IR materials and optical plastics discussed in this chapter.

10

The Visual
Optical System

10.1   Introduction

This chapter will be devoted to a description of the human eye that
will be adequate for purposes of understanding how the eye functions
and how it will interact with visual optical systems. Also covered will
be the topic of optical design and lens design, as they apply specifically to systems that are used with the eye as the final detector. Because
the human eye is a living and dynamic component, it is important that
the optical designer be aware of some details regarding its construction and function. Each individual is unique, in terms of specific eye
characteristics. As a result, the visual optical system must be designed
with all these potential differences in mind.

10.2   Structure of the Eye

Figure 10-1 shows a cross section view through the typical human eye.
This is a simplified version, suitable for the purpose of understanding
the basic function and makeup of the eye. The eye is nearly spherical
in shape, with a diameter of approximately 1 in. Because it is filled
with a water like substance, the eye is flexible (similar in consistency
to a tennis ball). The front of the eye contains the cornea, a convex
transparent window about 12 mm in diameter that allows light to
enter the eye. Because of the cornea's curvature, it is responsible for
most of the lens power that is present in the eye. The volume of the

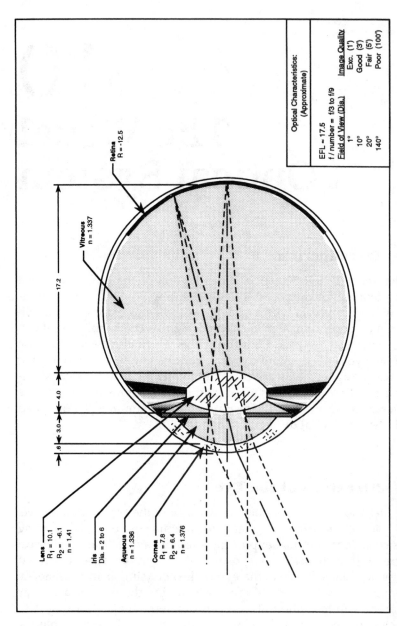

Figure 10-1. The human eye is a sophisticated optical instrument, quite similar in function to a camera. The representation above is a simplified version of the average eye. The actual eye is considerably more complex and will vary in its characteristics from person to person.

eye between the cornea and the lens is filled with a clear fluid called the *aqueous*. Just in front of the lens is the iris, an opaque layer with an opening at its center that will adjust in diameter depending on the amount of light that is present in the scene being viewed. The diameter of the iris opening will vary over a range of 2 to 6 mm. The normal size, for typical daylight viewing, is about 3 mm.

While the lens of the eye is quite powerful (EFL≈9.6 mm in air), it is considerably less powerful in its actual environment due to the fact that it is immersed in a fluid that has nearly the same index of refraction as the lens itself. The cornea is found to contribute about 60 percent of the total lens power of the eye (34 diopters), while the lens contributes the remaining 40 percent (23 diopters). A key function of the lens is its ability to change shape and, as a result, its lens power, thus allowing the eye to focus on objects at distances from infinity down to 10 in. This ability of the eye to change focus is referred to as accommodation.

The lens of the eye is followed by a relatively large distance to the focal surface of the eye, which contains the retina. The volume between the lens and retina is filled with a clear, waterlike fluid called the *vitreous*. The *retina*, like the film in a camera, is the light-sensitive surface on which the image of the eye is formed. The retina is a concave surface, with a radius of approximately 13 mm. Components of the retina are capable of converting the image information to electrical signals which are transported to the brain via the optic nerve.

Figure 10-2 shows the approximate field of view for each eye and the combined field of both. While the active area of the retina is large enough to give each eye a 140°+ field of view, the area over which we can resolve fine detail is limited to just a few degrees. Whenever an object of interest is detected near the outer edges of the field of view, the head and eye will be quickly rotated to bring that object to the center of the eye's field of view for closer scrutiny. Figure 10-3 illustrates the rapid loss of visual acuity as a function of field angle.

The curvature of the outer surface of the cornea is the key factor responsible for correct vision. When that curvature is too little or too great, the result is hyperopia (farsightedness) or myopia (nearsightedness). Astigmatism results when the cornea's surface is toroidal rather than spherical. These vision defects can generally be easily and effectively corrected by eyeglasses or contact lenses. Recently, surgical procedures have been developed to modify the shape of the cornea's outside surface to correct certain vision defects. With age, the flexibility of the eye's internal lens becomes less and the amount of accommodation (focus) that is possible is reduced and eventually (around age 60),

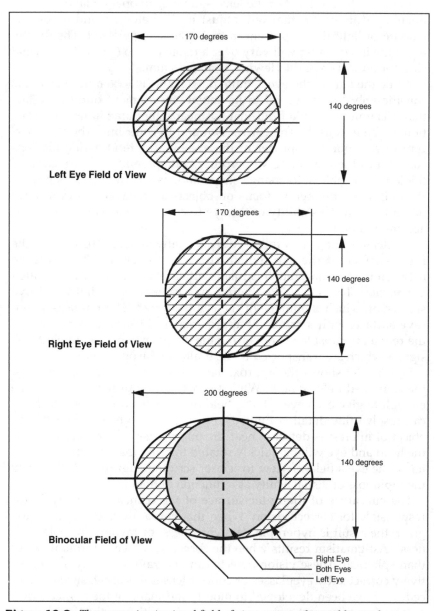

Figure 10-2. The approximate visual field of view, monocular and binocular.

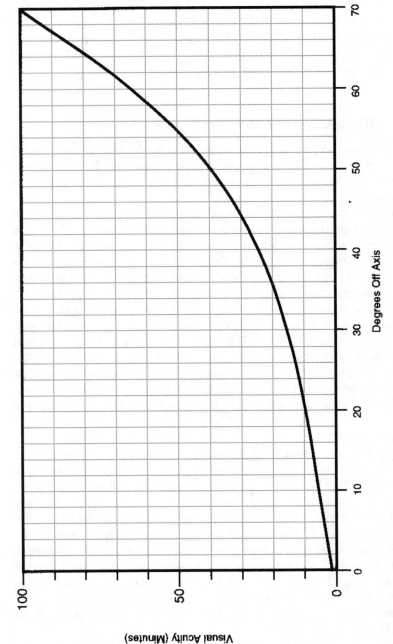

Figure 10-3. The visual acuity (resolution) of the eye is very much dependent on the location of the image on the retina. Maximum acuity is achieved over a very small field, limited by the size of the eye's fovea.

eliminated entirely. A more serious problem, also associated with aging, involves cataracts that cloud up the lens, reducing its transmission and causing scattered light that impairs vision severely. Here again, relatively straightforward procedures have been developed where the lens (with cataract) can be surgically removed and replaced with an implanted artificial lens to restore normal vision.

10.3 Resolution of the Eye

This section will cover, in greater detail, the resolution or visual acuity of the typical eye. The discussion will be limited to the central few degrees of the field of view, where the ability to resolve detail is at its maximum.

In the most common test for visual acuity the observer is placed at a known distance (usually 20 ft) from a chart containing several rows of block letters, each row of letters being progressively smaller in size. A person with *normal* vision will be able to resolve those letters that subtend a vertical angle of 5 minutes of arc. At 20 ft, a letter that is 0.35 in (8.9 mm) in height will subtend that 5-minute angle (see Fig. 10-4).

In the field of optical engineering, it is more common to designate resolving power in terms of cycles per millimeter or, for angular resolution, cycles per milliradian. The block letter of the visual test chart may be thought of as being made up of five horizontal strokes, or $2\frac{1}{2}$ cycles of a repeating pattern of equal-thickness black and white bars. For the example shown in Fig. 10-4, the spatial frequency of the target letter is 2.5 cycles/8.9 mm = 0.28 cycles/mm. The angular frequency of this same letter, at a distance of 20 ft, is 2.5 cycles/1.45 mrad = 1.72 cycles per mrad.

By applying the established principles of thin lens analysis and the known optical characteristics of the eye, we can construct the example illustrated in Fig. 10-4. Here we have a visual resolution (acuity) test chart at a distance of 20 ft and a *normal* viewer, whose eye is forming an image of that chart on the retina. As we can see from the figure, the image is inverted. Fortunately, the image processing portion of the brain is able to deal with this and we perceive the scene as being erect. The geometry of this example leads to the conclusion that the image of the 8.9-mm-high block letter E will be about 0.025 mm, or 0.001 inches high on the retina. In the more familiar optical design terms, this converts to a visual resolution capability of 1.72 cycles/mrad or 98 cycles/mm at the eye's retina.

Figure 10-5 contains an aerial image modulation (AIM) curve for the retina and MTF curves for the optics of the eye (pupil diameter = 3

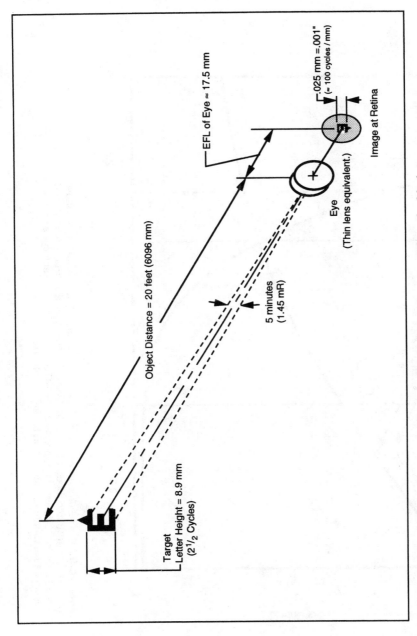

Figure 10-4. Conditions of the standard eye test for visual acuity. When the block letter that is 8.9 mm high is resolved by the eye at a distance of 20 ft, the viewer is said to have normal, or 20/20 vision.

241

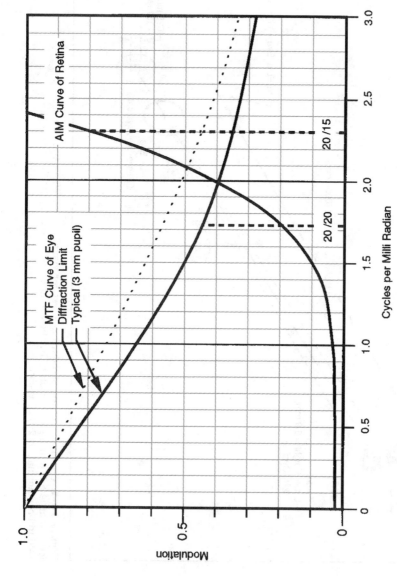

Figure 10-5. The resolution of the eye can be represented in several ways. Shown here is an aerial image modulation (AIM) curve for the typical retina, under normal viewing conditions. When combined with the MTF curve of the average eye's optics, it is possible to estimate the resolution of the combination.

mm) when it is diffraction limited and, more typically, when there is a small amount of image degradation. The AIM curve indicates the modulation that must be present in the image at the retina in order for the pattern to be resolved. The MTF curve indicates the modulation that will be produced by the optics of the eye for that image. These curves represent typical data and will tend to vary slightly from person to person. From Fig. 10-5 it can be seen that at a frequency of 1.72 cycles/mrad which corresponds to the standard (20/20) visual acuity test target, the retina requires a modulation of 0.20, while the eye's optics are delivering an image with a modulation of 0.45. The result is that the 20/20 target will be resolved. The next line of letters on the visual acuity chart is labeled 20/15 and contains letters that are smaller by a factor of 0.75×, resulting in an angular frequency of 2.3 cycles/mrad (at a distance of 20 ft). At this frequency it can be seen from Fig. 10-5 that the retina requires a modulation of 0.80 for resolution, while the eye's optics are delivering an image with a modulation of only 0.35. The result is the 20/15 target will not be resolved. It is the goal of the optical engineer, when designing an optical instrument to be used with the eye, that the degradation of the eye's MTF, due to the instrument, be kept to a minimum.

10.4 Visual Instrument Design Considerations

An example of the procedure used to generate the design of a basic visual instrument will be helpful in illustrating a few of the major considerations involved. First, it is essential that the actual application and requirements of the instrument be clearly defined. For this example we will assume a hypothetical case where it is required that we be able to visually observe and record motor vehicle license plate numbers from a distance of 100 yd. Starting data indicate that the height of such numbers (or letters) is typically 3 in, and that each figure can be assumed to consist of five horizontal strokes, similar to the visual acuity test target described earlier. While it is perfectly logical and correct to describe the target distance in yards and the letter height in inches, the first step in setting up any optical design problem for analysis is to convert all dimensions to the same units, in this case we will use millimeters. From Fig. 10-6 we can see that the typical target letter will subtend an angle of 76.2/91440 = 0.83 mrad. It was determined earlier that in order to visually resolve this type of target it must subtend an angle of at least 1.45 mrad. This confirms the need for an optical

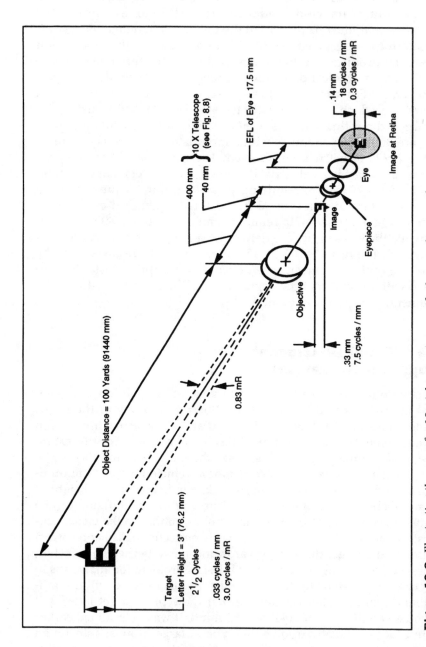

Figure 10-6. Illustrating the use of a 10× telescope to magnify the angular size of a target, such that it is possible to visually resolve 3-in-high letters at a distance of 100 yards.

Object Distance = 100 Yards (91440 mm)

10 X Telescope (see Fig. 8.8)

400 mm
40 mm

EFL of Eye = 17.5 mm

.14 mm
18 cycles / mm
0.3 cycles / mR

Image at Retina

Eye

Eyepiece

Image

Objective

.33 mm
7.5 cycles / mm

0.83 mR

Target
Letter Height = 3" (76.2 mm)
2¹/₂ Cycles

.033 cycles / mm
3.0 cycles / mR

instrument to magnify the apparent size of the target, in order to observe and resolve the required detail at that target. The introduction of a telescope with a magnification of 2× would theoretically produce an image that would be adequate for these purposes. However, if no other obvious negative factors arise, it would be prudent to consider a telescope with greater magnification to assure the required performance. In an earlier chapter a 10× telescope was described that would seem appropriate to this application. The basic optical parameters of that telescope (taken from Fig. 8-8) have been added to the layout in Fig. 10-6. It can be seen that two images of the target are formed in this system, one within the telescope and the final image on the eye's retina. It is useful to calculate the image size and spatial frequency at both image planes. Based on earlier analysis, we might conclude that if the spatial frequency of the image at the retina is less than 100 cycles/mm, or the angular frequency of the target to the eye is less than 1.72 cycles/mrad, then that target will be resolved. This assumes that the MTF of the eye's optics has not been significantly degraded by the introduction of the optical instrument. In reality, we are introducing an optical assembly between the eye and the target which we must assume contains some amount of optical error. Experience indicates that the amount of error present in a precision optical instrument such as this will be in the order of $\frac{1}{2}$ wave of optical path difference (OPD).

Let's digress here for a moment to consider the effect that small amounts of wavefront error will have on the MTF of an optical system. Figure 10-7 shows a normalized MTF curve for a diffraction-limited system and curves for systems of the same f number and wavelength, when small amounts of wavefront error are present. It is clear from this illustration why a system with $\frac{1}{4}$ wave of error is considered to be essentially diffraction-limited. It can also be seen that for an error of $\frac{1}{2}$ wave, the MTF cutoff frequency (maximum resolution) is about 0.5 of the diffraction limit, while a wavefront error of 1 wave reduces the cutoff to about 0.2, or one fifth of the diffraction limit.

The corresponding diffraction limit for the typical eye, under standard viewing conditions, with a 3-mm pupil, is about 300 cycles/mm, which converts to 5.25 cycles/mrad. It is reasonable to assume that the optics of this typical eye might introduce about $\frac{1}{4}\lambda$ of OPD error. From Fig. 10-7 it can be seen that at 0.5 of the cutoff frequency, a diffraction-limited system will have a modulation of 0.4, while a system with $\frac{1}{4}\lambda$ error will have a modulation of 0.3. Figure 10-8 shows again the AIM curve for the retina and the MTF curves for a typical eye along with one that has acquired an additional $\frac{1}{2}\lambda$ OPD error due to the introduc-

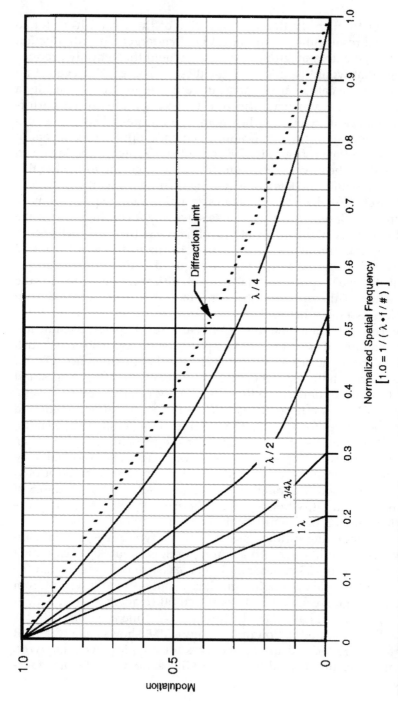

Figure 10-7. Normalized MTF curves for an optical system, showing the reduction in MTF due to small amounts of wavefront error in the system.

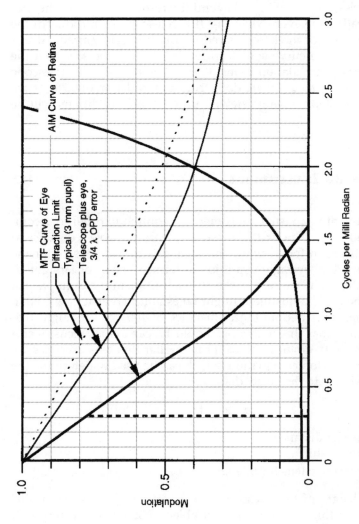

Figure 10-8. The resolution of the eye will be degraded by the addition of an optical system, in this case a 10× telescope (see Fig. 10-6). While the visual MTF is degraded, as shown by the ¾λ curve, the angular frequency of the target has been reduced from 3.0 to 0.3 cycles per milliradian. As the curves show, at that frequency (0.3) the required image modulation at the retina is just 0.03, while the optics of the telescope plus eye will deliver a modulation of 0.78, making the target easily resolvable.

247

tion of the 10× telescope. The critical point to note here is that, while the MTF has been degraded, the telescope has, at the same time, increased the apparent size of the letters being viewed, such that their angular frequency is now 0.3 cycles/mrad. Looking back at Fig. 10-6, we can see that the angular frequency of the target letters to the naked eye is 3.0 cycles/mrad, which is beyond the resolution limit of the retina (≈ 2.4/mrad). However, when the 10× telescope is introduced, the angular frequency of the letters is reduced to 0.3 cycles/mrad. At this frequency, from the curve shown in Fig. 10-8, it can be seen that the retina requires a modulation of just 0.03, while the eye plus telescope combination (the $\frac{3}{4}\lambda$ curve) are delivering a modulation of 0.78. As we anticipated, the letters being viewed under these conditions will be easily resolved.

10.5   Visual Instrument Focus

Focus adjustment in a visual instrument is required in order to accomplish two purposes. The first of these is to vary the object distance at which objects are in focus, while the second is to compensate for the viewer's individual eyesight condition in terms of near- or farsightedness. For instance, in the case of the 10× telescope in the example just described, the focus adjustment might be established as follows. First, the application specified an object distance of 100 yd. We might reasonably assume that this telescope should be capable of focus over a range from infinity down to 25 yd. From earlier thin lens exercises we will recall the newtonian equation, which can be used to determine the focus shift relative to the object distance:

$$x \cdot x' = f^2$$

where x = object distance
   x' = focus shift
   f = objective lens focal length

The focal length of the telescope's objective lens in this case is 400 mm. Using the thin-lens formula, an object distance of infinity will result in a focus shift of zero. For the specified viewing distance of 100 yd (91,440 mm), the focus shift will be 1.75 mm and, for the minimum object distance of 25 yd (22,860 mm), the focus shift will be 7.0 mm. Figure 10-9 (top) shows the image plane location for each of these cases. If the telescope were designed such that the eyepiece could focus

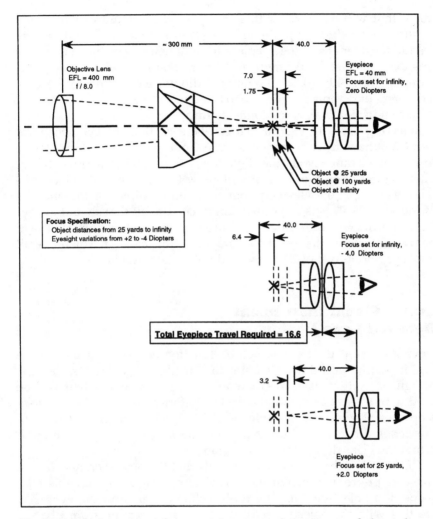

Figure 10-9. Instrument focus must accommodate a range of object distances and also, eyesight variations that may exist among those individuals using the instrument.

on the infinity image plane and then be adjusted (away from the objective) over a distance of 7.0 mm, this would accomplish the goal of being able to focus the telescope on objects from infinity, down to 25 yd.

The eyepiece adjustment to compensate for viewer differences is generally described in terms of diopters. While specifications will vary, a typical requirement will call for an instrument that is capable of being adjusted over a range of from −4 to +2 diopters. This is the range that will be used for this example. The amount of lens travel required to produce a 1.0-diopter focus shift is equal to $f^2 \div 1000$, where f is the focal length of the eyepiece in millimeters. For the eyepiece in this example the focal length is 40 mm, resulting in a focus travel requirement of $40^2 \div 1000 = 1.6$ mm per diopter. For a range of -4 to $+2$ diopters the eyepiece will have to be adjusted from -6.4 mm, to $+3.2$ mm from its nominal, zero diopter setting.

Referring again to Fig. 10-9, it can be seen that the eyepiece focal point will have to vary from a point 6.4 mm inside the infinity focus, to a point 3.2 mm beyond the 25-yd focus, for a total travel of 16.6 mm. This amount of eyepiece travel will allow for the change in object distance as well as the viewer accommodation required for the majority of individuals. This is a specification (focus travel) that should always be set on the generous side, to ensure that the basic requirement will be met with some degree of over travel. In this case, an actual overall total eyepiece travel of 20 mm would not be unreasonable.

10.6 Visual Instrument Detailed Lens Design

Having covered the basic structure, function and performance of the eye, it will be useful at this point to consider the details of the lens design procedure used to create an optical instrument that will be used with the eye as the final detector. Referring back to the 10× telescope described in Chap. 8 (Fig. 8-8), this section will present some of the detailed considerations that might be involved in generating an actual final lens design for this telescope.

There will be three major components that the optical engineer must consider in order to produce a complete optical design for this telescope: the objective lens, the roof-Pechan prism, and the eyepiece. Starting with the objective lens, the basic specification from Chap. 8 calls for a lens diameter of 50 mm, a focal length of 400 mm, and a 5° field of view. For a 10× telescope, this will result in a 50° apparent field of view at the eye. While this wide field of view is good for certain applications, such as sporting events and bird watching, it is a wider field than is required for a typical surveillance instrument. The design will be simplified by assuming that the field of view for this example will be 4° (40° at the eye). Assembly and alignment procedures will be simplified by the use of a cemented achromatic doublet lens form. The starting lens form might be selected from any one of numerous sources. In Chap. 7, a lens of these basic specifications was presented (see Fig. 7-5). We will use that lens as a starting point for this design. The correction of this lens will need to be modified, to account for the prism that is to be inserted into its back-focus region. Also, considering the glass types used in this lens, it will be noted that the original flint glass type is F7 (625356). This glass does not appear on the glassmap of select glasses that was shown in Fig. 9-1. It will benefit the design if we substitute F1 (626357), essentially an identical glass which does appear on the select glassmap. While this is admittedly a small change, it is typical of a change that is easy to make at this point in the design procedure and one that will have a positive impact on the ultimate producibility of the final system.

In order to produce a telescope with correct image orientation, it will be necessary to place an image erecting prism behind the objective lens. If we assume a roof-Pechan prism with a 40-mm aperture, the glass path through the prism will be $6 \times 40 = 240$ mm long.

The objective lens design will be reoptimized to include this 240-mm glass path, using the lens curvatures and focus as variables. The primary design constraints that must be controlled during the optimization procedure are a focal length of 400 mm and the correction (or balance) of spherical aberration and chromatic aberration to produce essentially diffraction-limited performance on-axis. The off-axis aberrations will be monitored and kept under control, but they will definitely assume a secondary role during the design process.

As the starting doublet design is shown in Fig. 7-5, it has an on-axis blur circle radius of 0.003 mm, indicative of essentially diffraction-limited performance (the Airy disk radius is 0.0055 mm). Changing the flint glass to F1 has essentially no impact on the performance of the lens. Adding the 240-mm thickness of BK7 to simulate the prism does disturb the state of correction, particularly the on-axis color, resulting in a blur circle with a radius of 0.012 mm, about twice the diffraction limit and an unacceptable value for this application. The doublet design is reoptimized to produce a lens-prism combination that is free of spherical aberration, primary axial color, and coma. Aberration curves for the resulting final design are shown in Fig. 10-10. It is important to keep in mind that the image produced by this lens will be viewed by the eye, using a 40-mm-focal-length ($\approx 6\times$) eyepiece. Examining the aberration curves, we see that the on-axis (FOB 0) curves are close to zero and, from the previous analysis, we can conclude diffraction-limited. As we go off-axis, we see the curves tilting, indicating a focus error and separating into three parallel curves, indicating off-axis, or lateral color. It is at this point that the fact that we are designing a visual system becomes significant. The focus problem, for example, will not be a problem if it can be easily accommodated (refocused) by the eye. The astigmatism, or field curves, in Fig. 10-10 show us that the maximum amount of focus

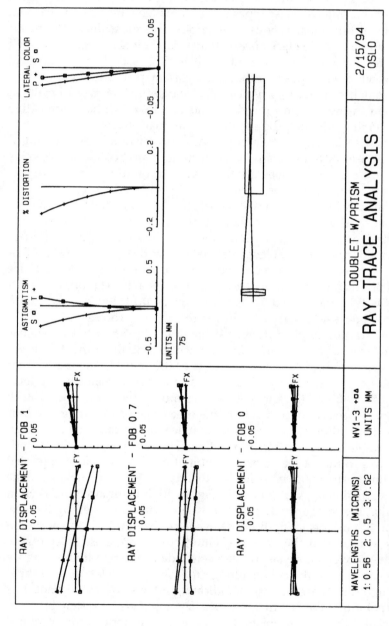

Figure 10-10. Aberration curves and lens drawing for a 400-mm, $f/8$ achromatic doublet, modified to operate with a 40-mm aperture roof-Pechan prism in its back focus to provide an erect image.

error is 0.8 mm for the tangential field and 0.4 mm for the sagittal field. From our earlier calculation of 1.6 mm per diopter for the 40-mm eyepiece being used, these numbers represent focus errors of 0.5 and 0.25 diopter, respectively. This amount of accommodation will not be a problem for a normal, healthy eye. The curves for lateral color show us that the maximum amount present is about 0.022 mm. When viewed through the 40 mm eyepiece, this represents an angular blur of $0.022 \div 40 = 0.0006$, or about 2 minutes of arc to the eye. This is an acceptable amount of lateral color that will not be noticeable when viewing most objects. The final curve in the ray trace analysis figure is the distortion curve. At less than 0.02 percent, distortion is obviously not a problem. The end product of this design exercise is a cemented doublet lens with a focal length of 400 mm, a speed of $f/8$, capable of imaging a 4° circular field of view, with an image quality that is quite acceptable for the 10× visual telescope application being considered. This lens will perform as predicted only when it is used in conjunction with a prism made from BK7 optical glass with a total internal path length of about 240 mm.

The prism design will be based on the standard roof prism configuration, as shown in Fig. 10-11. In addition to detailed manufacturing tolerances, the prism drawing should carefully spell out the overall optical quality required by specifying the maximum allowed wavefront error for a transmitted beam equal in diameter to the prism aperture. In this case a specification of ¼ wave single pass wavefront error, for a 40-mm-diameter bundle, will assure performance that is consistent with our telescope requirements.

The final optical component of the telescope is the 40-mm-focal-length eyepiece. There are many sources for eyepiece designs, including *MIL Handbook 141*, patent literature, and several textbooks. An eyepiece design that should work well in this application can be found in *Modern Lens Design* by Warren Smith (page 104). The design, as given in the book, has a focal length of 100 mm. This lens data can be taken into the computer and scaled to the required focal length using a subroutine that is a part of the OSLO lens design package. The resulting lens data for the eyepiece can then be merged with the objective lens design just generated, to form the complete 10× telescope. Fine tuning of the lens data includes adjusting the prism to eyepiece distance for best focus and sizing the eyepiece element diameters to ensure that the full off-axis light bundles are transmitted. The telescope analysis allows us to locate the system exit pupil, assuming the entrance pupil to be in coincidence with the first surface of the objective lens. The resulting eye relief is found to be 30 mm, a comfortable dimension for viewers with or without eyeglasses.

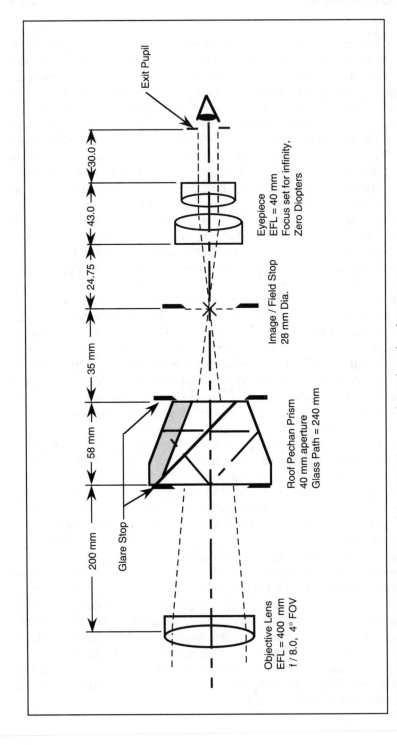

Figure 10-11. The optical design of the 10× telescope shown here has been generated using the OSLO lens design package. The resulting configuration is compatible with the function and performance of the eye.

The mechanical design of this telescope will include a 28-mm-diameter field stop at the internal image plane, along with several glare stops, or baffles, to eliminate stray light from outside the 4° field of view that might enter the objective and find its way to the eye. A convenient location for these baffles would be at the entrance and exit faces of the roof-Pechan prism. This choice of location will make it possible for the baffles to be made an integral part of the prism's mechanical mount. Figure 10-11 shows the final telescope design configuration. Computer analysis of this design shows it to be diffraction-limited on-axis, with about 2 diopters of astigmatism and 4 minutes of lateral color at the maximum field angle (these are combined objective plus eyepiece aberrations). The eyepiece also introduces about 5 percent distortion. All these values are indicative of a visual system of high quality and good optical performance. Figure 10-12 contains tabulated data on this telescope design, as generated by the OSLO Series 2 optical design software package.

10.7 Review and Summary

The material contained in this chapter has been presented to introduce the reader to the basic characteristics of the human eye, as they might affect the optical design of instruments intended for visual applications. The structure of the eye has been described and an optical model presented that illustrates the form and function of the individual components of the eye. The optical quality of the eye's optics and the retina were covered so that the resolution capability of the typical eye could be understood. The goal in describing the eye and its performance has been to assist the optical engineer in creating designs that are compatible with the eye. All data presented dealing with the eye has been classified as typical, taken from several sources and simplified to facilitate this analysis. Variations between individuals will be significant, making a more precise description inappropriate.

The final sections of the chapter have been devoted to a general discussion of the optical design of visual instruments, followed by a detailed lens design procedure as it might be used to generate the optical design of a 10× visual telescope. Analysis of that design has been reviewed, with particular attention to the unique aspects of systems intended for visual use.

```
*LENS DATA
10X TELESCOPE
SRF      RADIUS      THICKNESS    APERTURE RADIUS    GLASS   SPE   NOTES
1         --            --          25.00000 S        AIR

2      263.43000      8.00000       26.00000 A        BK7  C
3     -161.22000      6.00000       26.00000          F1   C
4     -510.85000    200.00000       26.00000          AIR

5         --        240.00000       25.00000          BK7  C
6         --         35.00000       25.00000          AIR

7         --         24.75000       13.99240 S        AIR

8         --          4.00000       20.00000          SF5  C
9       40.35000     20.00000       20.00000          SK11 C
10     -40.35000      2.00000       20.00000          AIR

11      36.95000     13.00000       18.00000          SK11 C
12     -38.47000      4.00000       18.00000          SF5  C
13        --         30.00000       18.00000          AIR

14        --            --           3.56169 S        AIR
```

```
*PARAXIAL TRACE
SRF     PY           PU           PI           PYC         PUC         PIC
1      25.00000    3.1250E-20   3.1250E-20     --          0.03492     0.03492

2      25.00000    -0.03239      0.09490       --          0.02300     0.03492
3      24.74092    -0.01978     -0.18585      0.18403      0.02152     0.02186
4      24.62226    -0.06250     -0.06797      0.31316      0.03466     0.02091

5      12.12238    -0.04117     -0.06250      7.24548      0.02283     0.03466
6       2.24126    -0.06250     -0.04117     12.72546      0.03466     0.02283

7       0.05378    -0.06250     -0.06250     13.93862      0.03466     0.03466

8      -1.49308    -0.03729     -0.06250     14.79649      0.02068     0.03466
9      -1.64226    -0.04280     -0.07799     14.87922      0.04820     0.38944
10     -2.49835    -0.03200      0.01911     15.84318     -0.14650    -0.34445

11     -2.56235     0.00460     -0.10135     15.55018     -0.24557     0.27434
12     -2.50258    1.6599E-06    0.06965     12.35772     -0.20817    -0.56680
13     -2.50257    2.7817E-06   1.6599E-06   11.52504     -0.34886    -0.20817

14     -2.50249    2.7817E-06   2.7817E-06    1.05920     -0.34886    -0.34886
```

Figure 10-12. Tabulated output data describing the 10× telescope design (Fig. 10-11), including basic lens data and paraxial ray trace data.

11

Lens Design and Image Evaluation

11.1 Introduction

This chapter will introduce the reader to the process of lens design and image evaluation. While not everyone working in the field of optical engineering will be called on to execute detailed lens designs, a basic understanding of the procedures involved, and their limitations, will be helpful to all. In the area of image evaluation, some degree of familiarity with the various methods employed will allow the optical engineer to more effectively specify required performance and later, to determine whether those levels of performance have been met. This chapter is particularly relevant when we consider the state of the computer and optical software field today. The days when access to these tools of lens design were limited to a very carefully chosen few are gone forever. In Chap. 5 we introduced the OSLO MG software package, pointing out its ready availability and ease of use in generating and evaluating certain (relatively simple) optical systems. This chapter will expand into the use of a more sophisticated software package (OSLO Series 2), outlining its use in generating detailed lens design solutions for a variety of specific problems. The field of optical design software is broad and very dynamic. There are a number of very fine packages to select from and they are all being constantly updated to reflect the state of the art.

11.2 The Lens Design Process

While there should always be flexibility in the process of generating a final lens design, there are a number of steps that are usually involved. These steps are listed below. In the lens design examples that follow, these steps will be followed and described in more detail.

The major steps in the lens design process are

- Define the problem (in optical terms) and establish requirements.
- Select a lens design to be used as a starting point.
- Modify the starting design, such that it meets all basic requirements.
- Optimize the modified starting design such that it conforms to general requirements (EFL, etc.) while providing the best possible image quality.
- Evaluate performance and compare with requirements; if good enough, go on.
- Document the final design (with tolerances).

11.3 10× Telescope Design

Defining the Problem

While it may sound trite, the definition phase of the lens design process is critical. Often the ultimate user of the lens system (the customer) can define the problem to be solved only in general terms, terms that are not consistent with those found in a true optical specification. The optical engineer must then work with the customer in generating a specification that all will be comfortable with. For example, in the last chapter we discussed a situation where the customer needed to be able to read the numbers on a license plate at a distance of 100 yd. The problem was that the customer could not resolve those numbers with the naked eye. The optical engineer's task in this case would be to come up with the concept of the 10× telescope (see Sec. 10.6), and to demonstrate to the customer how the use of this telescope would solve the basic problem. Having agreed on an approach, it is then possible to generate a lens design specification that will describe in detail the optical performance requirements of that telescope. The lens designer would then be tasked with generating the final lens design, including a tolerance analysis, optical schematic, and detail drawings of the components, suitable for their manufacture.

Selecting a Starting Point

Once the lens designer has received a complete description of the optical system requirements, the next step is to select a basic optical design configuration that will serve as a good and logical starting point for the lens design process. Past experience is perhaps the lens designer's most valuable asset at this stage of the lens design process. In addition, many of today's reference textbooks and optical design software packages will contain a library of basic lens designs. The combination of experience and access to a comprehensive lens design library will usually allow the designer to settle on a starting point quite quickly.

In the example of the 10× telescope, the lens design process can be broken down into two separate steps: the objective lens design (with prism), followed by the selection of an eyepiece design. In this section we will describe in more detail the procedures that the lens designer might have followed in generating the final telescope design. First, it was established that the objective lens should have a 400 mm focal length, a speed of $f/8$, and image a 4° field of view. A cemented doublet lens described in an earlier chapter of this book was found to meet these basic requirements and was selected as a starting point.

Modifying the Starting Point

The starting-point design was examined to determine any basic changes that might be made so that it was more suitable to the requirements of this particular design. In this case two changes were introduced, the flint glass type was modified to improve producibility, and a block of BK7 optical glass was added to the back focal region of the lens to simulate the required roof-Pechan prism.

Optimizing the Modified Starting Design

At this stage of the telescope design, the lens designer set the lens curvatures and the focus position as variables and used the optimization routine in OSLO to correct the basic aberrations of the objective lens.

Figure 11-1 illustrates the start and finish points of the objective lens optimization process.

The second step in the telescope design involves the eyepiece. It had been established earlier that an eyepiece would be required that was capable of covering a 40° field of view, with a focal length of 40 mm and a speed of $f/8$. A design was taken from a reference text and scaled to the required focal length. Its performance was found to be quite acceptable when combined with the objective lens and prism.

This is a classic example of the use of an existing lens design when possible. In this case, the objective lens design called for the presence of a large prism in its back focal space, making the use of an existing doublet design impossible. In the case of the eyepiece, the specifications and the method in which it was to be used were quite conventional, making it possible to take an existing eyepiece design and simply scale its dimensions to the required focal length and field angle.

Evaluating Performance

With the new objective lens and eyepiece combined to make a 10× telescope, the system was analyzed for image quality. It had been established that the primary goal of this design was to provide a telescope that would introduce no more than $\frac{1}{2}$ wave of error to the on-axis wavefront. One of the easiest and most effective methods of evaluating whether this goal has been met is to execute an MTF analysis which includes chromatic (color) and diffraction effects. This was done for this telescope design, the resulting MTF curve is shown in Fig. 11-2. The near coincidence of the actual MTF curve to the diffraction-limit curve shows that the wavefront error through the telescope is considerably less than $\frac{1}{4}$ wave. This would lead to the conclusion that, as required, the $\frac{1}{2}$ wave tolerance can be assigned to the various manufacturing aspects of the design.

Documenting the Final Design

Once a lens design has been generated and found to meet all system requirements, it is the lens designer's final responsibility to document that design thoroughly for purposes of manufacture and for future reference. For most new optical designs the following documentation should be provided:

- Tabulated lens data (prescription)
- Optical schematic

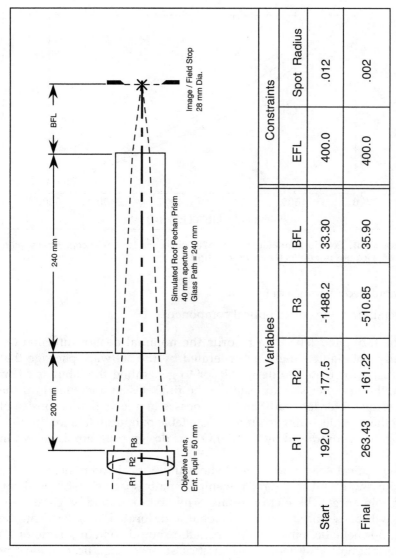

Figure 11-1. The objective lens for a 10 × telescope is shown with the basic variables and constraints that were used for its optimization.

	Variables				Constraints		
	R1	R2	R3	BFL	EFL	Spot Radius
Start	192.0	-177.5	-1488.2	33.30	400.0	.012
Final	263.43	-161.22	-510.85	35.90	400.0	.002

Objective Lens,
Ent. Pupil = 50 mm

Simulated Roof Pechan Prism
40 mm aperture
Glass Path = 240 mm

Image / Field Stop
28 mm Dia.

200 mm

240 mm

BFL

R1 R2 R3

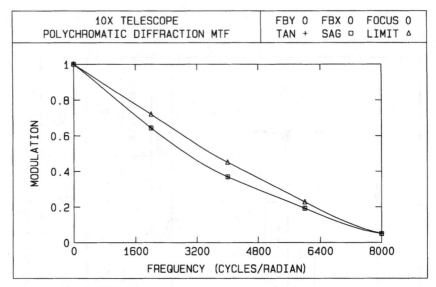

Figure 11-2. Polychromatic diffraction MTF curve for 10 × telescope final design, on-axis with 50-mm diameter entrance pupil.

- Tabulated tolerance data
- Detail drawings of all optical components

The tabulated lens data records the nominal design information. Ideally, this table should be generated by the software package that was used to create the design. Prior to generating the tabulated lens data, the designer should examine the final design and round off the values given for lens radii and thicknesses to a level that is consistent with the manufacturing tolerances. The table of optical data for the 10 × telescope, as produced by the OSLO lens design program, is shown in Fig. 11-3a.

The optical schematic should show the optical components of the system and establish the relationship (usually spacing) between them, with tolerances where appropriate. Any aperture, field or glare stops that will be required, should be identified, located, and sized on the optical schematic. Finally, the schematic should contain a table of all important optical system characteristics, such as magnification, field of view, and light transmission. The optical schematic should be a comprehensive reference document, containing any and all optical data that will be frequently referenced during the existence of the optical instrument that has been created. Figure 11-3b is an optical schematic of the 10× telescope, illustrating the typical design data and other information that might be included.

```
*LENS DATA
10X TELESCOPE
SRF        RADIUS      THICKNESS    APERTURE RADIUS   GLASS  SPE  NOTES
1          --          --           25.00000 S        AIR

2          263.43000   8.00000      26.00000 A        BK7  C
3          -161.22000  6.00000      26.00000          F1   C
4          -510.85000  200.00000    26.00000          AIR

5          --          240.00000    25.00000          BK7  C
6          --          35.00000     25.00000          AIR

7          --          24.75000     13.99240 S        AIR

8          --          4.00000      20.00000          SF5  C
9          40.35000    20.00000     20.00000          SK11 C
10         -40.35000   2.00000      20.00000          AIR

11         36.95000    13.00000     18.00000          SK11 C
12         -38.47000   4.00000      18.00000          SF5  C
13         --          30.00000     18.00000          AIR

14         --          --           3.56169 S         AIR

*GENERAL DATA
OSLO 2.3 WLKA
    EPR         OBY           THO          CVO        CCO          UNITS
 25.00000  -2.7937E+19   8.0000E+20       --         --         1.00000
    IMS      AST      RFS        AFO       AMO      DESIGNER     IDNBR
    14        2        2          1        ANG        OSLO         631

*WAVELENGTHS
CURRENT WV1/WW1    WV2/WW2    WV3/WW3     WV4/WW4     WV5/WW5    WV6/WW6
1        0.56000   0.50000    0.62000     0.43584     0.70652    0.40466
         1.00000   0.50000    0.50000      --          --         --

*REFRACTIVE INDICES
SRF   GLASS      RN1/RN4      RN2/RN5      RN3/RN6      VNBR
1     AIR        --           --           --           --
2     BK7        1.51803      1.52141      1.51554      88.17003
3     F1         1.62853      1.63606      1.62323      48.99609
4     AIR        --           --           --           --
5     BK7        1.51803      1.52141      1.51554      88.17003
6     AIR        --           --           --           --
7     AIR        --           --           --           --
8     SF5        1.67585      1.68485      1.66956      44.20636
9     SK11       1.56526      1.56917      1.56240      83.48822
10    AIR        --           --           --           --
11    SK11       1.56526      1.56917      1.56240      83.48822
12    SF5        1.67585      1.68485      1.66956      44.20636
13    AIR        --           --           --           --
14    AIR        --           --           --           --
```

Figure 11-3a. Tabulated lens data for the 10× telescope, including all nominal information required for generating manufacture and assembly drawings.

11.4 Precision Collimator Lens Design

Defining the Problem

This example will deal with the design of an objective lens for a precision collimator that is to be used as part of a spectrometer system. It is the function of this collimator to collect radiation from a point source and project that energy in the form of a collimated beam. The basic

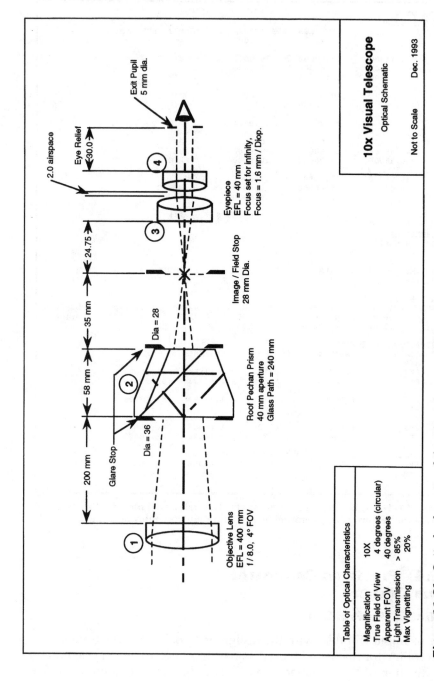

Figure 11-3b. Optical schematic of the 10× telescope, showing the information that might typically be included.

parameters established for this lens are a focal length of 300 mm, a speed of $f/6$ (or faster), and a spectral bandwidth of 0.6 to 1.1 µm. Because it is being used to create a well corrected, collimated beam, only the on-axis performance of this lens needs to be considered. The lens specification further states that overall image quality equivalent to less than $\frac{1}{2}\lambda$ wavefront error is required. The unique parameter of this lens, dictating the need for a custom lens design, is clearly the relatively broad, near-infrared spectral bandwidth.

Selecting a Starting Point

At this stage of the design process it is wise for the lens designer to consider the lens requirements in general and to prioritize them in terms of how each might impact the basic lens design. In this case the lens focal length and speed are not unusual and the field of view to be covered is essentially zero. It is probably safe to say that if all the lens requirements were this benign, a stock lens would have been selected and the lens designer's services would not be required. The spectral bandwidth, coupled with the need for near-diffraction-limited image quality, are the factors that will make this a challenging lens design problem.

For any lens or optical system, the spectral bandwidth is determined by the source, the detector, and the transmittance of all that lies between the two. The system engineer, responsible for the overall design of the spectrometer, might typically provide the lens designer with a spectral response curve similar to that shown in Fig. 11-4. The lens design must be optimized to transmit and image this bandwidth. For purposes of design and analysis the lens designer first divides this spectral band into several nearly equal bandwidth segments, assigning a wavelength and a weight to each segment. As the figure shows, the wavelengths selected for this design will be 0.85, 0.70, and 1.0 µm, with respective weights of 1.0, 0.9, and 0.9. The spectral weight is based essentially on the area under the spectral response curve for each segment.

Because of the relatively wide spectral bandwidth, it was concluded that either an all reflecting lens, or an apochromatic (corrected for three wavelengths) lens design would be required. Problems with physical layout and potential for alignment difficulties led to early elimination of the reflecting lens approach from consideration. The need for an apochromatic design was confirmed by calculations based on the graph that was presented earlier in Fig. 6-9. Since the lens speed

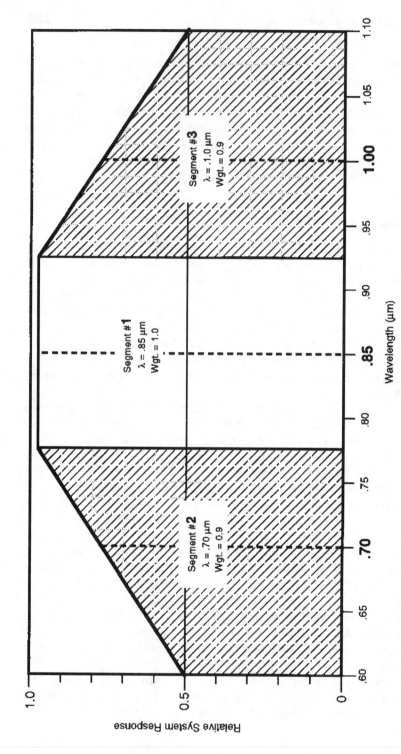

Figure 11-4. System spectral response curve, based on source output, detector sensitivity, and general system transmittance. Three design wavelengths and weights (based on area) are chosen as shown.

of $f/6$ given in the specification is presented as a minimum, we will enter this design process with a goal of $f/5$ for the lens speed (lens diameter of 60 mm). From Fig. 11-4 we see that the lens will operate at a central wavelength of 0.85 μm, with a spectral bandwidth of ± 150 nm. Having established these inputs, we can use Fig. 6-9 to estimate the blur spot radius due to secondary color in an achromatic lens form at approximately 0.015 mm.

The diffraction-limited blur circle radius (the Airy disk) for an $f/5$ lens, at λ = 0.85 μm, will be

$$R = 1.22 \cdot \lambda \cdot f \text{ number}$$

$$= 1.22 \cdot 0.00085 \cdot 5 = 0.005 \text{ mm}$$

Having a secondary color blur spot that is three times larger than the Airy disk tells us that the achromatic lens form will not perform adequately and that an apochromatic design will be required.

Like most refracting telescope designs, collimator objectives frequently take the form of doublet lenses. In this case, the need for a relatively fast lens suggests the need for an air-spaced doublet to permit adequate correction of spherical aberration. In Chap. 5 (Fig. 5-14) we described an interesting new design for an air-spaced, apochromatic, diffraction-limited doublet lens. This lens has been selected as the starting point for this design.

Modifying the Starting Point

The starting-lens design will have to be modified in several key ways if it is to be used for our collimator objective. First, the wavelengths for the design must be changed from the original visual band, to the values of 0.85, 0.7, and 1.0 μm. Next, the entrance pupil radius for the lens is set to 30 mm. Finally, the last surface of the lens is assigned paraxial solves for its curvature and for the distance to the image plane. The curvature is set so that the angle of the emerging paraxial marginal ray is −0.10, and the spacing is set so that the height of that ray is zero on the next (image) surface. These changes result in a lens that meets the basic system requirements of 300-mm EFL and a speed of $f/5.0$. The only remaining problem is image quality.

A quick look at the modified starting design indicates the blur spot radius for this lens form to be 0.17 mm, more than 30 times greater than the size of the Airy disk (our design goal for this lens). The optimization stage will tell us whether it is possible to manipulate this basic lens form such that it will deliver the required image quality.

Optimizing the Modified Starting Design

The optimization process involves the manipulation of available variables while monitoring the performance characteristics of the lens, particularly its image quality. For the apochromatic doublet, the four-lens surface curvatures and the focus position were used as variables. Constraints assigned during optimization held the focal length to 300 mm, while minimizing the on-axis chromatic spot size. After several cycles of the lens design process, an optimized configuration for this lens was generated. Figure 11-5 shows the start and finish configurations for this lens. The optimization of this doublet lens was accomplished by using the variable radii to alter the power of the individual elements while maintaining the sum of the powers constant (EFL = 300 mm). At the same time, the shapes of the elements were varied, along with the choice of the final image plane location (focus). The spot size that has been achieved indicates that an acceptable design has been generated. More thorough image evaluation will be done in the next stage of the design process to confirm this.

Evaluating Performance

The performance evaluation will cover two very basic and equally important areas. First, the basic lens parameters, including manufacturing tolerances and other aspects of producibility, must be considered. Second is the matter of image quality. Any design that measures up in one of these areas and falls short in the other must be considered a failed lens design effort. A lens that is easy to make and cheap, with poor image quality, represents a poor design. Likewise, a lens with theoretically superb image quality, that cannot be produced within established budget and schedule constraints, cannot be considered acceptable.

The apochromatic doublet design presents several manufacturing challenges that cannot be ignored. The most obvious of these is the choice of optical glass type for the positive element. In order to achieve the required apochromatic state of correction in a doublet lens form it is essential that at least one unusual glass type be used. In this case the positive (crown) element is made from the Schott glass type

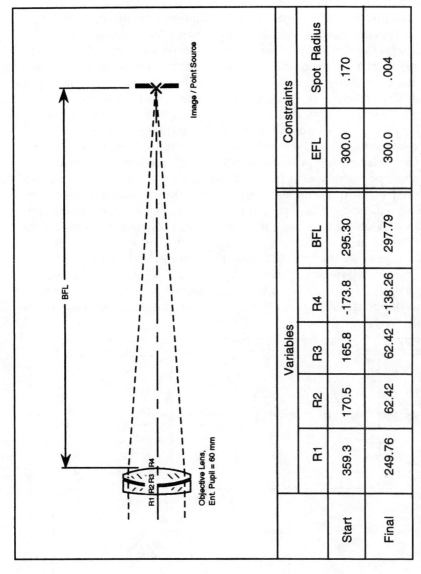

Figure 11-5. The objective lens for a near-IR collimator is shown with the basic variables and constraints that were used for its optimization.

| | Variables | | | | | Constraints | |
	R1	R2	R3	R4	BFL	EFL	Spot Radius
Start	359.3	170.5	165.8	-173.8	295.30	300.0	.170
Final	249.76	62.42	62.42	-138.26	297.79	300.0	.004

designated FK54 (437904). This glass is relatively expensive, not readily available, and quite difficult to work with because of an unusually high thermal expansion coefficient. On the other hand, the design requires just a single element of this material, weighing just a few ounces (optical glass is sold by weight), and the introduction of this glass into the design has been crucial to its success. The thermal expansion might be a problem were the instrument to be used under extreme environmental conditions, such as those that might be found in military or space applications. Fortunately, this is not the case for the spectrometer in question. The mechanical designer responsible for the lens cell design must take the glass characteristics into account while doing the design work. All in all, the tradeoffs in favor of using this particular glass type in this application would seem to prevail.

Optical performance, in terms of spot size, had been monitored and reported on throughout the design optimization phase. Now is the time for a more rigorous analysis of image quality. First, the aberration curves will tell the lens designer a great deal about the residual aberrations in the design and how the various aberrations have been balanced to produce the reported spot size. Figure 11-6 (upper left) shows the on-axis ray-intercept curves for this lens at the three principal wavelengths. These curves allow the designer to estimate a blur spot radius of about 0.004 µm. In the upper right, the corresponding wave-

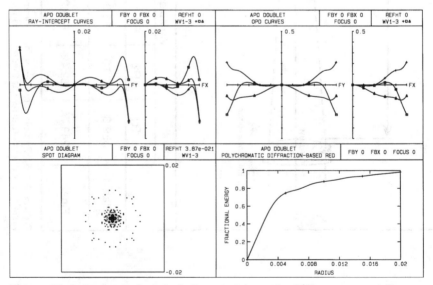

Figure 11-6. Image analysis including ray intercepts, OPD curves, spot diagram, and RED for a 300-mm, $f/5$, apochromatic doublet, designed to cover a broad spectral band in the near IR.

front (OPD) curves are shown. Here, OPD errors of about $\frac{1}{4}\lambda$ indicate the required near-diffraction-limited performance. The lower left quadrant contains the spot diagram and the radial energy distribution (RED) curve is shown in the lower right. Recalling that the Airy disk radius for this lens was 0.005 mm, this spot diagram analysis indicates essentially diffraction-limited performance for this lens.

This level of performance is ultimately and conclusively confirmed by the polychromatic, diffraction MTF data that is shown in Fig. 11-7. Having achieved all the established performance goals, our next step in the design process is to document the final design for manufacture and for future reference.

Documenting the Final Design

The first step in the design documentation process is to generate a table of lens data. It will assure accuracy if this is done using the lens design software, at nearly the same time that the image quality evaluation is being done. Obvious, but very important, is the fact that this approach guarantees that a lens produced per the tabulated data will yield image quality that is consistent with results of the just completed analysis. The apochromatic doublet is completely described by the

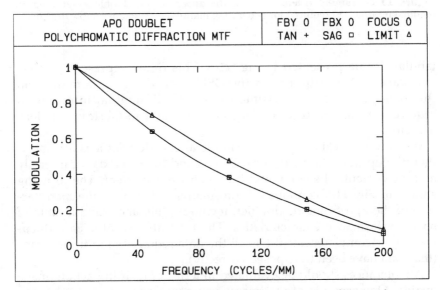

Figure 11-7. Polychromatic diffraction MTF for a 300-mm, $f/5$, apochromatic doublet, designed to cover a broad spectral band in the near IR.

```
*LENS DATA
APO DOUBLET
SRF       RADIUS      THICKNESS   APERTURE RADIUS   GLASS   SPE   NOTES
1          --            --          30.00000 A       AIR         Modeled glasses.

2       249.76000      8.00000       30.00000 S      BAK1 C
3        62.42000      1.00000       29.65427 S       AIR

4        62.42000     15.00000       29.85473 S      FK54 C
5      -138.26000    297.79000 V     29.78589 S       AIR

6          --            --           0.05672 S       AIR

*PARAXIAL CONSTANTS
        EFL          FNB          GIH          PIV         PTZRAD        TMAG
     300.00392     5.00007      0.05236     -0.00524    -371.71224   -1.3200E-10

*GENERAL DATA
OSLO 2.3 WLKA
        EPR          OBY          THO          CVO          CCO
     30.00000   -3.9667E+08   2.2727E+12       --           --         UNITS
        IMS     AST    RFS     AFO          AMO         DESIGNER     1.00000
         6       1      1       0           TRA                      IDNBR
                                                                       149

*WAVELENGTHS
CURRENT WV1/WW1     WV2/WW2      WV3/WW3      WV4/WW4      WV5/WW5      WV6/WW6
1         0.85000    0.70000      1.00000      0.43584      0.70652      0.40466
          1.00000    0.90000      0.90000        --           --           --

*REFRACTIVE INDICES
SRF    GLASS      RN1/RN4      RN2/RN5      RN3/RN6      VNBR
1      AIR          --           --           --          --
2      BAK1       1.56425      1.56798      1.56172      90.08126
3      AIR          --           --           --          --
4      FK54       1.43288      1.43477      1.43154     133.82310
5      AIR          --           --           --          --
6      AIR          --           --           --          --
```

Figure 11-8. Tabulated lens data for the apochromatic doublet, including all nominal information required for generating manufacture and assembly drawings.

tabulated data presented in Fig. 11-8. This figure represents typical lens data as it is output from the OSLO lens design program. From this table it should be possible to extract all nominal information required to manufacture the lens elements and to fabricate the lens assembly.

When the optical system is as simple as this doublet lens assembly, then the optical schematic, lens details, and tolerance data can easily and conveniently be combined into a single document. The drawing shown in Fig. 11-9 contains all information relevant to the manufacture of the apochromatic doublet, including tolerances and a table of basic performance characteristics. The ultimate test of such a document is its completeness (it has all the data required) and its accuracy (that data have been transposed correctly).

All tolerances should be evaluated in terms of their impact on image quality. One example of the tolerancing process will be given here to demonstrate this basic approach. The performance of the doublet lens

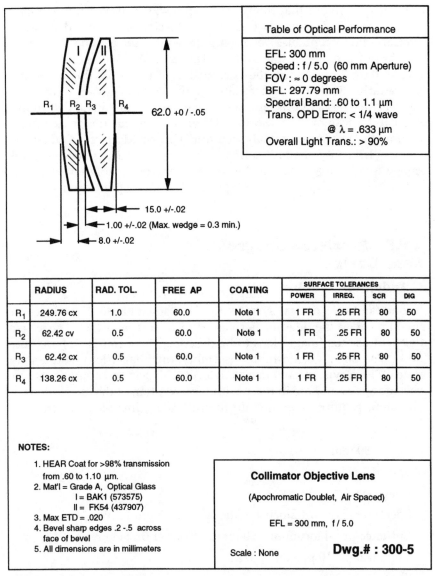

	RADIUS	RAD. TOL.	FREE AP	COATING	SURFACE TOLERANCES			
					POWER	IRREG.	SCR	DIG
R₁	249.76 cx	1.0	60.0	Note 1	1 FR	.25 FR	80	50
R₂	62.42 cv	0.5	60.0	Note 1	1 FR	.25 FR	80	50
R₃	62.42 cx	0.5	60.0	Note 1	1 FR	.25 FR	80	50
R₄	138.26 cx	0.5	60.0	Note 1	1 FR	.25 FR	80	50

Figure 11-9. Detail drawing of the apochromatic doublet, including all tolerance and performance information.

will be degraded when one of the elements is tilted relative to the other. The amount of allowable tilt must be determined and translated to the parallelism, or wedge tolerance on the lens spacer. Standard machining on a part of this type will easily produce parallelism within 0.001 in. Converting this to an angular tilt of the second element

(0.024°), it is found that the nominal spot size of 0.0043 mm increases to 0.013 mm. This is nearly three times the size of the airy disk, which is clearly not acceptable. Decreasing the angle between the elements to a tight but achievable tolerance of 0.0002 in (0.005°) results in a spot size of 0.0048 mm. Since the Airy disk size is 0.0053, this amount of degradation is deemed acceptable. Other tolerances on parameters such as thickness, radius, and wedge can be evaluated and determined in a similar fashion. For a simple design such as this it is reasonable to assume the tolerances will all accumulate in an additive, or worst-case fashion. In reality, the probability is great that the actual lens assembly image quality degradation due to manufacturing tolerances will be between one half and one quarter of the *worst-case* prediction.

11.5 Precision Imager Lens Design

Defining the Problem

In one family of optical systems the lens assembly that produces the final image on the detector is referred to as an *imager lens.* The system that will use the imager lens to be described here consists of an afocal unit (telescope), a beamsplitter assembly, the imager lens, and a solid-state CCD television camera as a detector. All system components have been designed and are in place with the exception of the imager lens.

System requirements dictate the following imager lens specifications:

EFL: 80 mm

Speed: $f/1.75$

Image size: 9.6 mm square

Spectral band: Monochromatic, $\lambda = 0.633\ \mu m$

Entrance pupil location: 60 mm in front of first element

Vignetting: 10 percent maximum

Spot size: 0.04 mm radius for 80 percent energy (all field points)

Selecting a Starting Point

Here again, at this stage of the design process it is wise for the lens designer to consider all the lens requirements and to prioritize them in terms of how each might impact the basic lens design. In this case the lens focal length and speed are not unusual. The field of view to be covered is reasonable at about ±5° maximum. The fact that this will be a monochromatic design will simplify it greatly. The requirement that the entrance pupil be located in front of the lens will make this design quite unique and difficult. The image quality specification is not severe, but the need for uniformity across the field, coupled with the relatively small vignetting allowance, will make the design a challenge.

It is usually a good idea at the start of each lens design to compute the Airy disk so that the image quality requirement can be compared with it. In this case, the diffraction-limited blur circle radius (the Airy disk) for an $f/1.75$ lens, at $\lambda = 0.633$ µm, will be

$$R = 1.22 \cdot \lambda \cdot f \text{ number}$$

$$= 1.22 \cdot 0.000633 \cdot 1.75 = 0.0014 \text{ mm}$$

Having a spot size requirement that is nearly 30 times the size of the Airy disk means that all design and analysis work can be conducted on a geometric basis, and diffraction and its effects can be ignored.

The starting lens configuration was chosen after several alternatives were considered and rejected. The Cooke triplet and double-gauss lens forms offer great promise in terms of lens speed, field of view, and image quality. However, the performance of both designs relies greatly on their basic symmetry about the aperture stop. For the imager lens, the entrance pupil is the aperture stop and, because it is located in front of the lens, that basic symmetry does not exist. A lens type that is frequently used with a forward entrance pupil is the Petzval type. It was concluded that a derivative of the basic Petzval lens form would offer a good probability of success in this case. A search of available literature turned up a four-element Petzval lens design that was selected as a starting point for this design.

Modifying the Starting Point

The starting Petzval lens design was then modified to conform to the basic requirements of focal length and lens speed that had been established. The optical glass types were simplified, to reduce cost and complexity (it was felt that the monochromatic nature of the design would permit this). It can be seen from Fig. 11-10 (left) that the upper ray of the off-axis bundle is being refracted through some severe

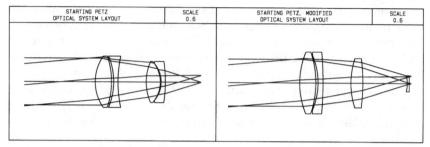

Figure 11-10. The requirements established for the imager lens indicate a Petzval lens form (left) for a starting point. That form was simplified to create the modified Petzval form shown at right, which was the form used as input to the optimization process.

angles and finally, at the last element, it is no longer transmitted through the lens. The steep inside curves of the rear doublet are obviously not desirable. It was surmised that these curves were involved in correcting color and field curvature or astigmatism in the original design. Since this is a monochromatic design, it was decided to replace the doublet with a single element and to add a field flattener close to the image plane to take care of the field curvature. This modified form of the starting Petzval lens was then subjected to a preliminary optimization, with encouraging results. That modified version of the starting Petzval design is shown in Fig. 11-10 (right).

This is a logical point in the lens design process to stop and review the design that is evolving in terms of its general configuration. In this case for example, it was confirmed that the presence of a field flattener lens in close proximity to the detector would not be a problem. The overall physical size of the lens (length and diameter) can be estimated with some certainty at this point. It should be confirmed that the size and weight of the proposed lens will be acceptable. A preliminary estimate of the approximate cost to manufacture the lens is also possible at this time, on the basis of the modified starting lens form.

Optimizing the Modified Starting Design

The final optimization process involves the manipulation of available variables while monitoring the performance characteristics of the lens, particularly image quality. For the imager lens, all surface curvatures and the two major airspaces were used as variables. The constraints assigned held the focal length to 80 mm, while minimizing the spot size for three field angles (0°, 3.5°, 5°). The optimization process depends on an effective software package, combined with its intelligent use by a skilled, experienced lens designer. It is a rare occasion when the designer can dump all requirements into the computer and produce a successful design with no further intervention. After several cycles of this design process, an optimized configuration for the imager lens was established. Figure 11-11 contains data describing the starting and finish configurations for this lens. The uniform spot size that has been achieved for all field points indicates that an acceptable design has been generated. More thorough image evaluation will be done in the next stage of the design process to confirm that this is indeed the case.

Evaluating the Design

As with earlier examples, the design evaluation covers two very basic and equally important areas. First, the basic lens parameters, including manufacturing tolerances and other aspects of producibility, must be examined. Second, the design's performance in terms of image quality must be evaluated. The design must meet all requirements and expectations in both areas in order to be considered a successful design.

The final imager lens design has a focal length of 80 mm, and a speed of $f/1.75$ and produces the required image quality over the entire image. The clear apertures of the lens elements were carefully assigned to assure that the vignetting specification would be met. All lens elements are made from common glass types, which were selected for their many desirable properties. A quick look at tolerance sensitivities indicates that the design should be quite reasonable as far as manufacture and assembly are concerned.

The image quality, in terms of spot size, had been monitored during the design optimization phase. Now is the time for a more rigorous analysis. First, the ray trace analysis will tell the lens designer a great deal about the residual aberrations in the design and how the various aberrations have been balanced to produce the reported spot size. Figure 11-12 shows the final aberration curves for this lens at the three field positions. These curves give a good indication of the uniformity of performance over the full field of view. Shown elsewhere are the field curves for astigmatism and distortion. The flat field due to the field lens is responsible in large part for the uniform image quality. The maximum distortion of 1 percent in the corners of the square image will be quite acceptable. The lens layout, including the marginal and chief rays, is helpful to the lens designer in evaluating the function and significance of the individual components.

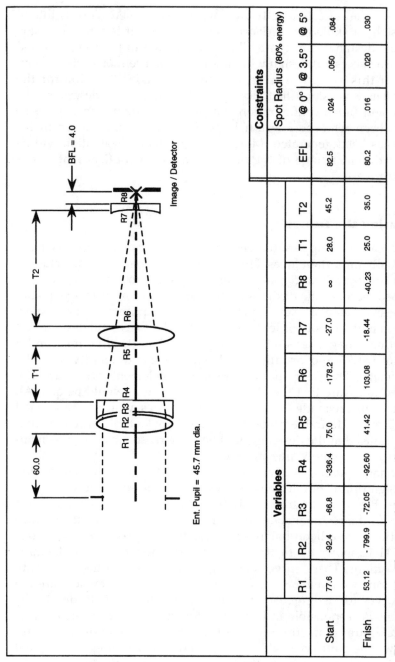

Figure 11-11. The optical layout of the imager lens is shown with the basic variables and constraints that were used for its optimization. Values are presented for the lens at the start and the end of the optimization process.

	Variables										Constraints			
													Spot Radius (80% energy)	
	R1	R2	R3	R4	R5	R6	R7	R8	T1	T2	EFL	@ 0°	@ 3.5°	@ 5°
Start	77.6	-92.4	-66.8	-336.4	75.0	-178.2	-27.0	∞	28.0	45.2	82.5	.024	.050	.084
Finish	53.12	-799.9	-72.05	-92.60	41.42	103.08	-18.44	-40.23	25.0	35.0	80.2	.016	.020	.030

Ent. Pupil = 45.7 mm dia.

BFL = 4.0

Image / Detector

R1 R2 R3 R4 R5 R6 R7 R8

60.0 T1 T2

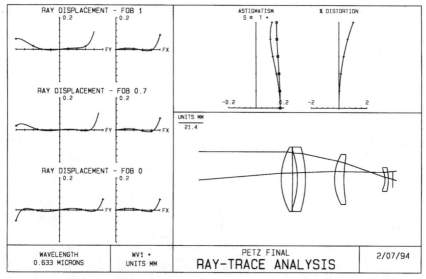

Figure 11-12. Ray trace analysis of the final imager lens design. This basic output from the OSLO Series 2 optical design package shows all aberration data along with a convenient lens layout.

Since the original specification of image quality is given in the form of spot size, and because diffraction effects are not a consideration, the ultimate evaluation for image quality of this lens is done using spot diagram analysis. Figure 11-13 shows the spot diagrams for the three field positions. Also shown are through-focus spot diagrams which indicate the sensitivity of focus for this design. Finally, the RED curves allow the designer to confirm that the specification for 80 percent of the energy to be within a 0.04 mm radius has been met comfortably at all field positions.

Documenting the Final Design

Figure 11-14 is a table of lens data for the final imager lens design. As this is a compound lens assembly, the design calls for an optical schematic, followed by a set of detail lens drawings, one for each optical element (as shown in Chap. 5, Fig. 5-5). The optical schematic for this lens is shown in Fig. 11-15. In a lens of this complexity it will be most effective to add a separate sheet to the schematic, containing a tabulation of all tolerance data relative to the assembly and alignment of the lens. The function of Fig. 11-15 then would be limited to illus-

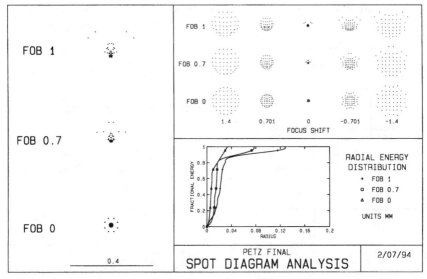

Figure 11-13. Spot diagram analysis of the final imager lens design. This basic output from the OSLO Series 2 optical design package shows spot size and shape, through-focus spot diagrams, and RED curves.

trating the nominal lens configuration and documenting the basic optical characteristics of the lens assembly.

11.6 An Unusual Lens Design Task

In nearly every case, it is the lens designer's task to generate a design that has well-corrected image quality. There is one unique category where the designer departs from this philosophy and instead generates a design with very carefully controlled image quality that is sometimes far from the usual goal of perfection. This case arises when designing a null lens for the fabrication and test of an aspheric surface. The following examples will illustrate the procedures used to generate two typical null lens designs.

First, consider the case where a primary mirror is to be produced for use in an IR telescope assembly. This mirror has a diameter of 260 mm, a vertex radius of 355 mm (concave), and a conic constant of -1.0828. A mirror with a conic constant of 0 will be spherical. When its conic constant is -1.00, it is a paraboloid, and when it is less than -1, as is this case, the mirror surface is a hyperboloid. An ideal method for testing a concave reflecting surface is to generate a monochromatic wave-

```
*LENS DATA
PETZ FINAL
SRF      RADIUS       THICKNESS    APERTURE RADIUS    GLASS   SPE   NOTES
1         --             --         22.91400           AIR

2         --          60.00000      23.00000 A         AIR

3       53.12000      10.00000      27.00000           BK7 C
4     -799.90000       4.00000      27.00000           AIR

5      -72.05000       7.00000      24.00000 K         BK7 C
6      -92.60000      25.00000      27.00000           AIR

7       41.42000       8.00000      21.00000           BK7 C
8      103.08000      35.00000      18.00000           AIR

9      -18.44000       3.00000       8.50000           F4  C
10     -40.23000       4.00000 V    10.00000           AIR

11        --             --          6.70704 S         BK7 C
12        --             --          6.70704 S         AIR

13        --             --          6.70704 S         AIR

*PARAXIAL CONSTANTS
      EFL          FNB          GIH          PIV         PTZRAD        TMAG
   80.26992     1.75155      6.67274     -1.90481    2097.99488  -1.0034E-19

*GENERAL DATA
OSLO 2.3 WLKA
      EPR          OBY          THO          CVO           CCO          UNITS
  22.91400  -6.6503E+19  8.0000E+20       --            --          1.00000
      IMS     AST     RFS     AFO          AMO       DESIGNER        IDNBR
      13       2       2       0           TRA                       1170

*WAVELENGTHS
CURRENT WV1/WW1     WV2/WW2      WV3/WW3      WV4/WW4      WV5/WW5      WV6/WW6
1       0.63300     0.70000      1.00000      0.43584      0.70652      0.40466
        1.00000       --           --           --           --           --

*REFRACTIVE INDICES
SRF  GLASS      RN1/RN4      RN2/RN5      RN3/RN6      VNBR
1    AIR          --           --           --          --
2    AIR          --           --           --          --
3    BK7        1.51508        --           --          --
4    AIR          --           --           --          --
5    BK7        1.51508        --           --          --
6    AIR          --           --           --          --
7    BK7        1.51508        --           --          --
8    AIR          --           --           --          --
9    F4         1.61313        --           --          --
10   AIR          --           --           --          --
11   BK7        1.51508        --           --          --
12   AIR          --           --           --          --
13   AIR          --           --           --          --
```

Figure 11-14. Tabulated lens data for the imager lens, including all nominal information required for generating manufacture and assembly drawings.

front that reflects from that nominal surface in a predictable fashion, and then to monitor that reflected wavefront, adjusting the shape of the surface until it performs as predicted. When the surface is spherical, the center of the projected wavefront is set to coincide with the center of curvature of the mirror and, when the mirror surface has been figured to the correct (spherical) shape, the wavefront will be

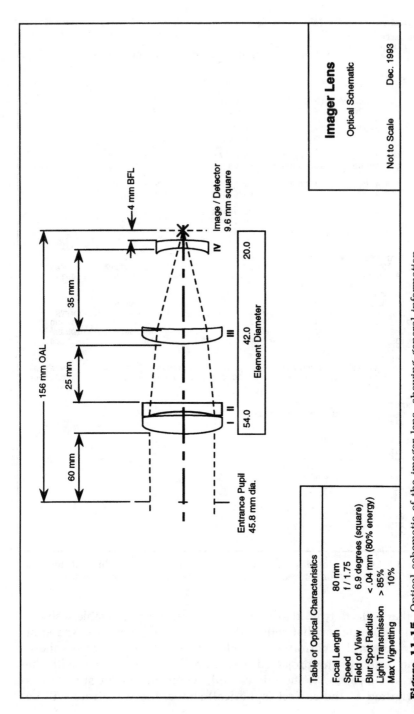

Figure 11-15. Optical schematic of the imager lens, showing general information that might typically be included. Comprehensive tolerance data would be contained on a second sheet. Elements I through IV would be described completely on separate detail drawings.

reflected precisely back on itself. This reflected wavefront can then be monitored interferometrically and the mirror surface corrected to a small fraction of a wavelength.

If the mirror under test is a paraboloid, then the center of the projected wavefront is placed at the focus of the mirror. The reflected beam will then take the form of a plano wavefront which can be reflected from a flat reference mirror, once again from the mirror under test and then evaluated interferometrically. Having a conic constant very close to −1.0, the mirror to be tested in this example is very nearly a paraboloid. If we wanted to test this mirror as a paraboloid, we would assume a point source at its focus and evaluate the error that exists in the reflected beam, which would be perfectly flat for a paraboloid. When the hyperboloid in question is thus evaluated, it is found that there is an OPD error of many wavelengths in the reflected wavefront. This would not be acceptable in this case, since we are striving for a test configuration with less than 1 wave of error.

Further analysis reveals that this hyperboloidal mirror has two foci, one at a point 174 mm in front of the mirror and a second that is located 8749 mm behind the mirror. The significance of these foci being that a wavefront originating at one of them, will be focused by the mirror at the second, with zero wavefront error. Knowing this, we can conclude that if we place our interferometric point source at the first focus, 174 mm in front of the lens, the reflected wavefront from a properly finished mirror will be perfectly spherical and it will appear to be coming from the second focus, located 8749 mm behind the surface. This condition is illustrated in Fig. 11-16. It can be seen that if a spherical test mirror is placed at the proper location, the matching spherical wavefront will be reflected back on itself, such that it precisely retraces its original path. With a test setup as shown, any error in the returning wavefront will be a precise indication of the error that exists in the hyperboloidal mirror under test. The spherical test mirror in this configuration is also known as a *Hindle sphere*.

The Hindle sphere is one of the simplest and most basic form of null lens. More typically, the aspheric surface will be more complex and the null lens design will be generated using a somewhat different approach. For example, the same IR telescope design that uses the primary mirror just described, has a concave aspheric surface in the form of an oblate spheroid with a vertex radius of 31.84 mm, a diameter of 28 mm, and a conic constant of 0.0806. The approach to a null lens for this surface is to generate a design that will produce a diverging wavefront that precisely matches the shape of the desired aspheric surface. The departure of the surface under test from spherical will be an indicator of the complexity that will be encountered in the null lens

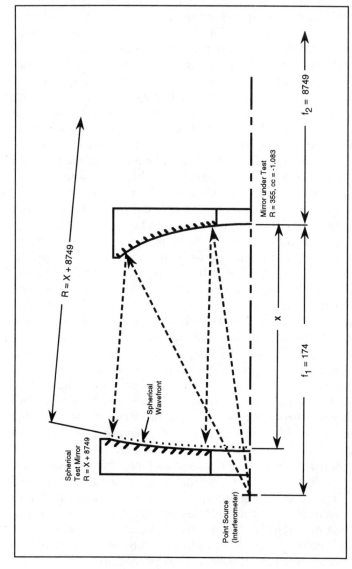

Figure 11-16. Test setup for manufacture and evaluation of a concave, hyperbolic primary mirror. When a point source is located at one focus of the hyperboloid f_1, the reflected wavefront appears to emerge from the second focus f_2. The spherical test mirror (null lens) is set with its center of curvature coincident with that second focus, making it possible to test the hyperbolic mirror interferometrically.

design. In this case, the surface asphericity is greater than five waves. Past experience indicates that the null lens design for manufacture and test of a surface having this much asphericity will take the form of a two-element lens assembly. Once the basic lens form is selected, the design is then optimized, using all lens curvatures as variables. Figure 11-17 shows the resulting arrangement, including the prescription of the null lens design. As with the Hindle sphere, the surface test is accomplished by introducing a point source which creates a perfect spherical diverging wavefront. In this case, the null lens introduces additional divergence, along with wavefront aberration that allows the transmitted wavefront to exactly match the surface under test. In terms of ray tracing, when the surface has reached the required shape, each ray emerging from the null lens strikes the surface under test at exactly a 0° angle of incidence. The result then is that each ray is then reflected back on itself, retracing its path through the null lens and back to the point source. The shape of the reflected wavefront is evaluated interferometrically, with any error being an indication of residual error in the surface under test.

These two examples of null lens design have been introduced to give the reader some feel for the diversity of work that might be done by the lens designer. The null lens falls into the category of test equipment and as such must be handled differently when it comes to tolerancing and documentation. The key point here is that the null lens will, in most cases, not be delivered to the customer but will rather become a tool in the inventory of the manufacturer. In general, basic tolerances for test equipment must be held tight enough to assure the performance of the lens being manufactured. Often the null lens tolerances will be tighter than those on the finished part by a factor of ten. On the other hand, cosmetic defects such as edge chips and scratches, and certain physical tolerances such as lens diameter, will not have to held as tight. An additional economic consideration is that a good null lens assembly will serve for the manufacture of many parts, thus its initial cost can be spread out over the life of a contract involving many repeat uses.

11.7 Review and Summary

This chapter has been presented to introduce the reader to the field of lens design and to demonstrate a few of the unique and interesting aspects of that area of optical engineering. Three essential factors are required to create a successful design: a skilled and experienced lens

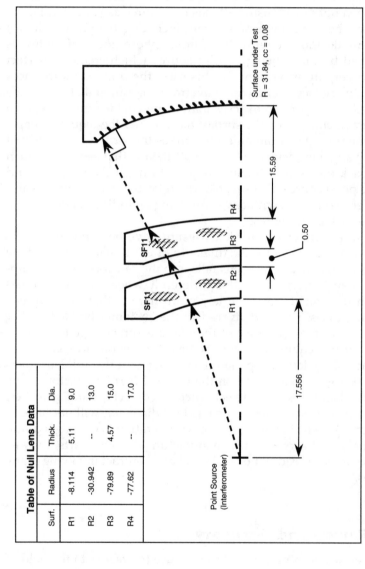

Table of Null Lens Data

Surf.	Radius	Thick.	Dia.
R1	-8.114	5.11	9.0
R2	-30.942	--	13.0
R3	-79.89	4.57	15.0
R4	-77.62	--	17.0

Figure 11-17. Test setup showing a two-element null lens for the manufacture and evaluation of a concave aspheric surface. When a point source is located at the point shown, the refracted wavefront emerging from the null lens will be shaped to exactly match the desired aspheric surface at the distance shown. Interferometric testing will reveal any error in the shape of that surface.

designer, a well-executed optical design software package (program), and a powerful and fast computer. While all three are essential, they have been listed in order of importance.

While the lens design process must remain flexible, a typical process has been outlined here that will serve well as a guide to one entering the field. Several examples have been presented to demonstrate the methods that might typically be employed in generating a lens design, using the suggested typical lens design process. The final section deals with the design of null optics for the fabrication and testing of aspheric surfaces. This has been included to demonstrate the wide range of design tasks that the lens designer might be called on to execute. In this case, the typical procedure is departed from in order to meet some rather unusual requirements.

All the work done in this chapter has been done using the OSLO Series 2 optical design program on a Gateway 486/33 personal computer. The field of computer hardware and software today is one of the most dynamic ever experienced in the history of technological development. While many changes will occur during the lifetime of this book, it is believed that this chapter will maintain its value as a tool to help the reader to understand and (hopefully) enjoy the subject of lens design.

12

Optics in Our World

12.1 Introduction

This book has revealed many facets of the field of optics and optical engineering. It has been my hope that this exposure will encourage some readers to further pursuits and studies, perhaps leading to a career in this field. For others this has merely been an excursion into an area of general interest and nothing more. I believe this final chapter will be of interest to all these readers. It will deal with the optical aspects of several phenomena and devices that we all might encounter in our daily lives. It is my hope that this will increase our awareness, appreciation, and understanding of them all.

We have seen in several earlier chapters how the human eye is closely related to the optical design of certain instruments. This chapter will touch on the concept of the complete visual system, including the eyes, the mind, and our lifetime of visual experiences. These experiences tend to *program* the mind, allowing it to interpret what is seen by the eye and imaged onto the retina.

Also in this chapter, we will deal with a number of technological advances of the day, with special emphasis on their optical content. In order to keep this presentation at a reasonable comfort level (for the author as well as the reader), this section will be limited to the most fundamental concepts and principles.

12.2 Optical Illusions: Size

Many times what we actually see, and what we think we see, are two very different matters. This results when the mind, based on earlier experience, adds information to that collected and delivered by the eye, leading us to a conclusion that may not always be correct. In Fig. 12-1, for example, the diagram on the left represents a common optical illusion known as the *Ponzi illusion*. Here we see two rectangular blocks (*A* and *B*) that are exactly the same size. It follows that the images of blocks *A* and *B* on our retina are also the same size. In spite of these facts, the other clues included in the figure lead us to conclude that the upper block is substantially larger than the lower block. The deception begins when our mind relates the two near vertical lines to a lifetime of experience dealing with conditions similar to standing and looking down a long, straight highway, or stretch of railroad tracks. On the basis of this experience, we conclude that the two lines in the figure are really parallel and that they appear to converge only because the tops of the lines are much farther from us than the bottoms. It is a small leap then to conclude that the upper block is also farther away from us than the bottom block. Now the mind goes into action; seeing two blocks that appear to be the same size, but believing one to be farther away from us than the other, then obviously the one that is farther away must be larger. This, then, two different-size blocks, is not what our eye sees, but rather it is what our mind perceives. This is an optical illusion.

The diagram on the right in Fig. 12-1 is a similar illusion. The perspective layout indicates the inside of a box, with the sides receding from us. The three arrows shown are all the same height. Because other clues indicate they are at different distances from us, we conclude that they are progressively larger as their distance from us increases. For most viewers, both of these illusions will be slightly enhanced when viewed with one eye rather than two. This happens because when we view an object with both eyes, the eyes converge to a point on that object. In the case of the arrows, this clue (the amount of eye convergence) tells our mind that all arrows are at the same distance. This tends to offset, to some degree, the other clues that are indicating otherwise. Viewing with just one eye removes the convergence clue from the set of information being processed by the mind.

The moon illusion is one that has been witnessed by most of us. When we see a full moon rising over the horizon, it is generally perceived to be much larger than when it is seen more nearly overhead. The basis for this illusion is quite similar to those just described and illustrated in Fig. 12-1. The concept of the celestial sphere involves an imaginary surface on which all the stars, planets, and other heavenly

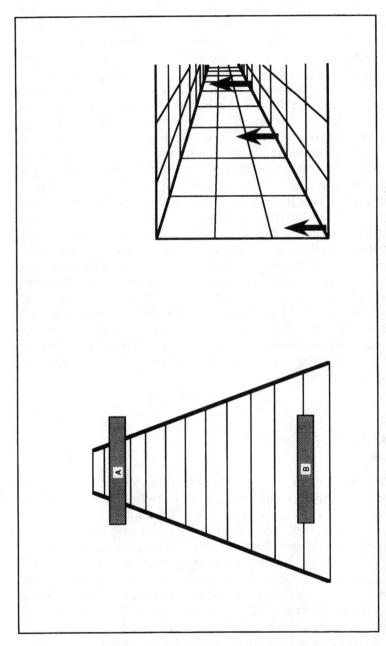

Figure 12-1. While the eye is a precise optical device, the image produced on the retina is interpreted by the mind, where other visual clues are introduced which sometimes fool the viewer into reaching the wrong conclusion. For example, is block A on the left really larger than block B? Are the three arrows in the box at right really different heights?

bodies are located. If we stand in the middle of a large flat expanse, on a clear night, it is not difficult to imagine such a surface. For most of us, as with the observer in Fig. 12-2, we perceive this celestial sphere to be flattened considerably relative to a true sphere. This may be due to our lifetime of exposure to cloud-filled skies during the daylight hours, where those clouds near the horizon are considerably farther away from us than those that are nearly overhead. In any event, as we gaze at the star-filled night sky, it is quite likely that the surface containing all the stars, planets, etc. will be perceived as having the shape shown in Fig. 12-2. As the moon rises above the horizon, it is assumed to be on that surface and its size is judged accordingly. As the night progresses and the moon appears more nearly overhead, our presumptions indicate that it is now closer to us. Because its angular size has remained constant, in spite of having (apparently) moved closer to us, we can only conclude that the moon has become smaller than it was when it was on the horizon. Again, our mind leads us to a conclusion that contradicts what our eye sees and what our intelligence tells us is actually the case. As in Fig. 12-1, where making actual measurements of the blocks and arrows to confirm their size does not cause the illusion to no longer be seen, knowing that the size of the moon is constant does not diminish the effect of the illusion. In these examples, as with the moon illusion, what we see, what we know, and what we think, are passed to the brain and processed to generate the ultimate perception of the mind.

12.3 Other Optical Illusions

Optical illusions such as the moon illusion relate to our judgment of the size of certain objects. Other illusions deal with the perception of shapes and objects that might not otherwise actually exist. For example, the diagram on the left in Fig. 12-3 at first appears to be just a collection of straight lines of different lengths, some horizontal, some vertical. The significance of the pattern is not obvious (to most of us) on the basis of our experience viewing similar patterns. When it is suggested that this might be a block letter *H* that is being illuminated from the upper left, our mind accepts that premise and the complete outline of the letter is now perceived. It will now be difficult to view this pattern of lines ever again without having the letter *H* appear.

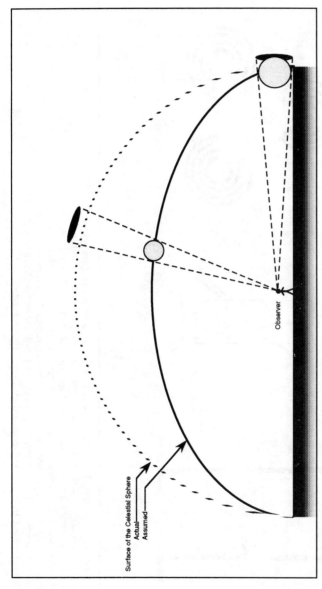

Figure 12-2. The moon illusion causes the apparent size of the moon to change, depending on its location in the sky. When the moon is near the horizon, we erroneously assume it to be at a greater distance than later in the night, when it is nearly overhead. Because we assume the celestial sphere to be flattened, we further assume that as the night progresses, and the moon travels away from the horizon, it is also moving closer to us. Like blocks *A* and *B* in Fig. 12-1, the size of the moon is the same in both cases. If the moon appears the same size, in spite of its having (apparently) moved closer to us, we can only conclude that it has grown smaller as it has moved from the horizon to the near overhead position.

Surface of the Celestial Sphere
Actual
Assumed

Observer

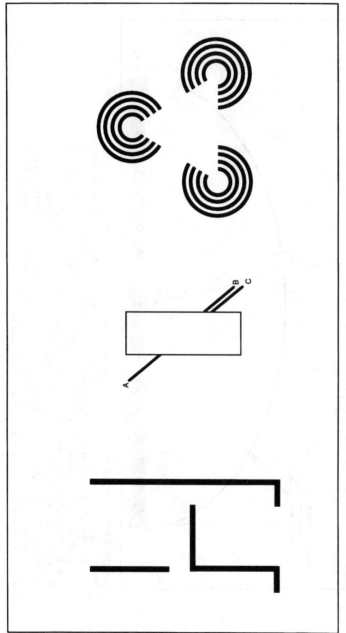

Figure 12-3. The pattern on the left appears at first to be a random set of straight lines until it is suggested that this might represent a block letter *H*, which is illuminated from the upper left. The lines are then interpreted as shadows of the letter edges and the letter *H* is easily perceived. In the center figure, which of the lines on the right (*b* or *C*) is an extension of line *A*? On the right, the arrangement of interrupted circles leads to the perception of a white triangle, in spite of the fact that there are no lines present to indicate the actual outline of that triangle.

In the center of Fig. 12-3 we can see a rectangular block with a diagonal line (A) passing behind it from left to right. Emerging on the right are two parallel lines, one of which is a true extension of line A. The interruption of the line, combined with its diagonal orientation, leads us to conclude that line B is the extension of A. If we lay a straight edge on the figure, it is apparent that our mind has again been fooled. Interestingly, this is another case where being made aware of the reality does not cause the illusion to go away.

While in the case of the letter H, it was apparent only after we had been given a suggestion of its existence, the diagram on the right in Fig. 12-3 requires no such clue. Here we have a pattern of incomplete circles from which a white triangle literally leaps out at us: this, in spite of the fact that no white triangle exists. Our mind leads us to conclude that the only reason these circles would appear as they do is because this white triangle has been placed between our eyes and the circles. So, the nonexistent white triangle is clearly perceived by our mind's eye.

12.4 Seeing the Third Dimension

As we view the outside world, our visual system easily perceives the three dimensions of width, height, and depth. When we view the printed page, the sensation of depth is greatly diminished. When we are viewing printed reproductions of the outside world, in the form of drawings and photographs, there are generally enough clues present to allow us to perceive depth where it does not really exist. Methods exist that allow us to maximize and even enhance the third dimension when actually viewing a two-dimensional object. In the mid-1800s the stereoscope became a familiar and popular method of home entertainment. Utilizing the basic optical arrangement shown in Fig. 12-4, the stereoscope simultaneously accomplishes three functions, each adding to the three-dimensional illusion. First, the lens power that is present allows the viewer to relax the focus of the eyes, which indicates to the mind that the scene being viewed is at a great distance. Second, the prism effect of the lens segments allows the viewer to relax the convergence of the eyes, again indicating a distant scene. Finally, and most important, the stereoscope is designed to view a pair of photos of the same object, that have been recorded from slightly different points of view, much as the object would have been seen

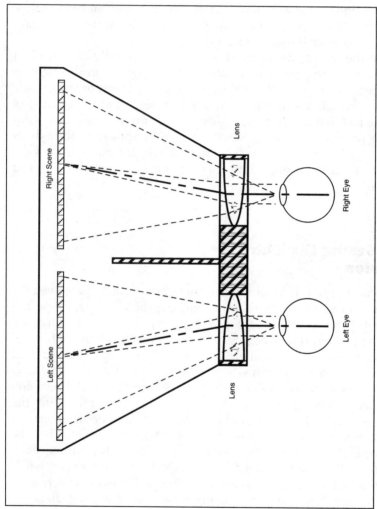

Figure 12-4. The stereoscope is a viewing device that produces a three-dimensional effect when it is used to view a properly prepared stereo pair of pictures. The lens segments introduce both lens and prism power, which allows the eyes to relax both focus and convergence. The final, and most important, function of the stereoscope is to direct the left scene to the left eye and the right scene to the right.

by the viewer's two eyes. The instrument is configured such that each eye sees one of these photos and not the other. The result is the eyes focus and convergence are relaxed, while each eye sees a view of the scene from a slightly different point of view. When this information is processed in the viewer's mind, the illusion of a true three-dimensional scene is realized. In modern time the original stereoscope has been repackaged into the familiar "viewmaster," an educational toy used for the viewing of a variety of cartoon characters as well as real-world scenes by our children.

A pseudo-stereo effect will be observed by most viewers by simply viewing a printed image with one eye while the other eye remains open but covered. This effect is most pronounced when viewing a large picture of a scene containing numerous clues relating to depth. An object such as a road, river, or fence that extends from close to the point of observation in the foreground, deep into the scene, will help reinforce the illusion. The result of covering one eye will generally be that the convergence of the eyes will relax and the axes of the two eyes will become nearly parallel. This removes the convergence clue from the process and allows the mind to accept the fact that objects in the scene may be at varying distances, relative to their appearance and size. The illusion of depth that will be achieved depends on the relaxed state of the viewer and the content of the image being viewed. Once achieved, it is interesting to uncover the second eye and note how the effect immediately ceases to exist following the introduction of the convergence clue.

Three-dimensional optical systems have come and gone from the entertainment scene over the years. The 3D movie came on the scene in the 1950s, but failed to catch on. In large part this was due to the advent of several spectacular wide-screen processes (Cinemascope, Cinerama, Todd AO, etc.). It was found that the added benefits of a wide-screen presentation were better received than the somewhat artificial effects of 3D viewing. In more recent years the advantages of 3D have come to the fields of tactical reconnaissance and have proved to be very worthwhile. In another related development, the endoscope used in noninvasive surgical procedures has recently been modified to provide a true three-dimensional view to the surgeon. The advantages in this application have been described as very significant in the success and effectiveness of such surgeries. While the potential for a home 3D TV system seems to exist, it appears at this time that in television, as it was with the motion picture, the large-screen approach will win out once again.

12.5 Optics and the Compact Disk

Many of the optically related technological advances of today have impacted the field of home entertainment. In just a few short years the compact disk (CD) has essentially replaced the phonograph record and magnetic tape as the primary sources for recorded music in the home and automobile. This has been made possible by simultaneous advances in laser systems and data-processing electronics. The phonograph record, in spite of many improvements over the years, has long been basically a spiral groove in a flat disk. With output from a microphone being used to record the sound, the configuration of the groove is physically altered. On playback, the needle (stylus), tracking in the groove of the record, will vibrate in a way that is analogous to the sound waves that were recorded. The mechanical motion of the stylus is then converted to an electrical signal, which is amplified and processed to drive a loud speaker, thus reproducing the original sound.

All this has been changed by the ability to produce a comparable digital electronic signal that contains all the information needed to reproduce the recorded sound. In the digital recording system the output from the microphone is converted into a continuous digital signal. This signal is then converted to the binary form that is used in the processing of most computer data. In the binary language, all numerical data are converted to a series of bits, each represented by either a zero (0) or a one (1). In this way the recorded signal can be processed using techniques developed for computer applications. The result is that any piece of recorded music (or other information) can be transformed into a long string of 0s and 1s. This string of binary digital data is then transferred to a spiral track on a CD, where one of the binary digits is made highly reflective while the other is made to absorb incident energy. Now, the optical aspect of the CD system comes into play. The CD is played back while being rotated at a variable rate, such that the linear speed of the data track in constant. Energy from a solid-state laser source is collected by a precision lens assembly and focused down onto the data track. The reflected laser energy is collected and analyzed, creating a digital electronic signal that precisely duplicates the original recorded data. The accuracy of the digital process is enhanced greatly by the fact that the playback process is noncontact; thus wear and tear on the record (CD) is minimal and its life is essentially infinite. Figure 12-5 illustrates the basic principle of the CD and the CD playback mechanism.

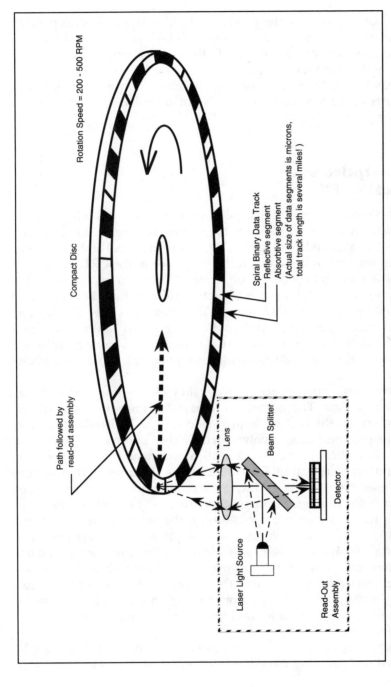

Figure 12-5. The compact disk (CD) playback mechanism involves a precision turntable, functioning in conjunction with a read-out assembly containing: a solid-state laser light source, a beamsplitter, a precision lens assembly, and a detector which senses the signal reflected from the data track on the disk.

A similar approach to the playback of video signals is incorporated into the *video laser disk* system, where both picture and sound are contained on a somewhat larger disk. Both the CD and the laser disk can replace audio and video magnetic tape. In addition to the advantages of increased fidelity and a reduced wear factor, the quick access to random parts of the recorded program is greatly enhanced by the digital approach.

12.6 Optics and Projection TV

The most common form of television set in use today is referred to as *direct view*. In the direct view set the picture is generated by electron guns within a cathode-ray tube (CRT, or picture tube) and the viewer observes the output directly, on the front face of the CRT. While still the system of choice based on overall image quality, the CRT approach is limited by the size of picture tube that can be reasonably produced and used in the home. In the quest for larger displays, at reasonable cost, several projection TV schemes have been developed that offer quite good performance today and considerable promise for the future. All these systems involve rather complex and sophisticated optics.

The two basic approaches to projection TV are the *front* and *rear projection systems*. The names are derived from which side of the viewing screen the picture is projected on. As is usually the case, there are pros and cons involved with either approach. Figure 12-6 shows schematically, the typical optical arrangement used for both front and rear projection TV systems. The front projection system shown uses three sources which generate identical pictures, except for their color content. Beamsplitters are used to combine the three pictures which are then projected onto the front surface of a large screen. As with the conventional motion picture, the projected scene is reflected from the screen for viewing by the audience This front projection approach offers a compact projection unit and considerable flexibility with regard to the ultimate size of the projected image. On the down side, the front projection system is made up of two separate units (projector and screen) and requires a clear line of sight between the two.

The rear projection system is somewhat more bulky, but it is a self-contained unit, offering more flexibility in terms of its location within

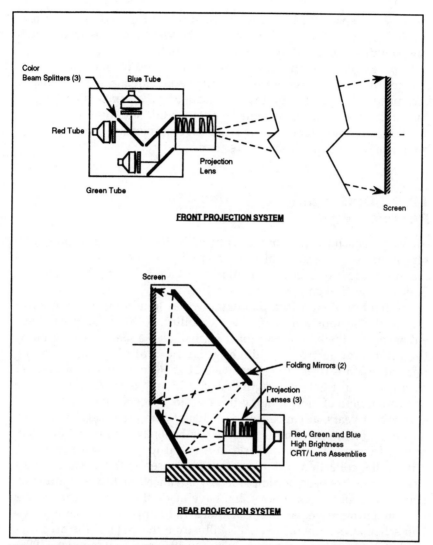

Figure 12-6. The conventional home TV uses a CRT, or picture tube, to present the picture to the viewer. Size and cost limitations of the CRT approach have led to the development of several projection TV systems. Shown here are two typical systems, illustrating the level of optical design and engineering that is involved.

ewing space. In this case the picture is projected onto the rear surface of the screen and it is seen by the viewer as it is transmitted (and slightly dispersed) by the screen material.

As programming sources become more numerous and the TV system continues to merge with computer and video game systems, it becomes more common to find a dedicated *media center* as an integral part of the design in many home layouts. As this situation continues to develop, it would seem that the large screen projection TV system will become more affordable and more commonplace.

12.7 Optics and Photography

As we approach the turn of the century, the field of amateur photography is undergoing several monumental changes. This section will touch on a few of these, particularly as they relate to the subject of optics and optical engineering.

The field of 35-mm film photography has long been the choice format of the serious amateur. The flexibility of this format has been enhanced by the availability of interchangeable lenses, which allow the photographer to vary the angular field of view that is being imaged. With a fixed film format of 24 × 36 mm, the photographer might select a lens focal length between 18 and 400 mm to produce a horizontal field of view ranging from 90° to 5°, respectively.

In recent years, advances in the fields of computerized lens design, lens manufacture, and antireflective coatings have led to the development of compact, high-quality, multifocal-length (zoom) lenses to replace the collection of interchangeable lenses in the 35-mm camera system. This has been accompanied by advanced methods of automatic exposure and focus control that have made the modern camera not only much more versatile but also nearly foolproof. The camera system schematics shown in Fig. 12-7 illustrate the state of the art in the field of 35-mm photography, as it was in 1970 and as it is today (1994). In the case of the modern camera, it is now possible to combine state-of-the-art optics with solid-state electronics and microprocessors, into a single compact package that has all the capabilities of earlier systems, and more. The apparent lack of available lens speed in the newer, more compact, cameras has been more than offset by recent developments in the field of high-quality, very fast, color print and slide films.

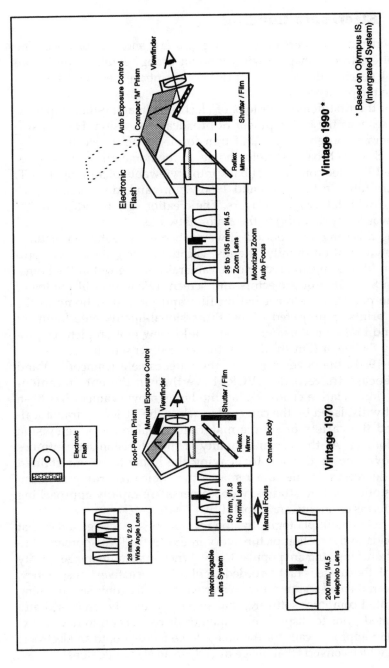

Figure 12-7. The 35-mm single-lens reflex (SLR) camera has long been the instrument of choice for the serious amateur photographer. The two camera systems illustrated show some of the significant advances that have been realized in recent years. Progress in the fields of lens design and optical engineering have contributed significantly to these advances.

12.8 Electronic Imaging

Developments in this area are coming at a fast and furious pace. The status of electronic imaging will no doubt change significantly before this book reaches the reader. A brief review of the field will serve to illustrate the point.

One of the first demonstrations of electronic imaging that impacted the masses was the development of broadcast television. Here, a studio camera recorded images and transformed them to electronic signals which were broadcast to the home. In the home, the television set converted the electronic signal back into a viewable picture on a CRT, or picture tube. In the 1950s and 1960s, more and more homes were equipped with television receivers. The creation and broadcast of TV signals was left (primarily) to the major networks.

During this same time period, if we wished to have motion pictures of our own creation (family, friends, vacations, etc.), these were produced on film, using amateur movie cameras and viewed in the home, using an 8- or 16-mm projectors and screen. Likewise, still photos of that time period were recorded on film and viewed at home in the form of prints or projected slides. Professional-quality entertainment was found in the theaters, where one could view motion pictures produced by the major film studios, on the large silver screen.

In the 1980s, the development of the videocamera (camcorder) and the videocassette recorder (VCR) have thrown the entertainment industry into relative chaos. Today the home movie camera has been essentially displaced by the camcorder, while the VCR has dramatically altered the variety of source material available to the television viewer forever. As the size and quality of the television set continues to improve, the motivation for leaving the home to be entertained rapidly decreases. While film remains the choice recording medium for the motion picture studios, videocameras are rapidly approaching them in terms of quality and capability.

In the field of still photography, the revolution is at a stage today that corresponds to the motion picture story in the 1970s. Two approaches to still electronic imaging are popular today. When maximum image quality is the goal, the image is first recorded on film. The negative, transparency, or print can then be scanned and converted to an electronic signal. Using sophisticated computer software, that signal can then be processed and manipulated prior to display on a monitor or conversion to hard copy. The second approach calls for the image to be focused onto an electronic detector, which converts that image directly to an electronic signal.

This entire topic (electronic imaging) represents an extremely beneficial fusion of multiple technologies. The camera, both still and motion picture, the projector, the screen, the print (or slide), scanners, the TV camera, TV set, camcorder, VCR, and the modern home computer system, have all come together in a spectacular blending of technologies. Optical engineering remains the keystone of the process, with a quality lens responsible for the capture of the original image and its transfer into the electronic domain. Computer hardware and software are major contributing factors to the success of any electronic imaging system, where the potential for image enhancement and manipulation appears to be essentially unlimited.

The ultimate in today's technology is perhaps demonstrated by the CD ROM (compact disk, read-only memory) system, which brings vast quantities of recorded pictures and sound to the computer, for instant access and display. These are indeed exciting times to be involved with the science of optics, as the wave of electronic imaging carries us on into the next century.

12.9 Optics of the Rainbow

Doubtless, one of the most beautiful and spectacular optical phenomena of nature would have to be the rainbow. Here we have, on a most grand scale, a demonstration of the lens, the prism, refraction, reflection, and dispersion, all being observed by the human visual system. Figure 12-8 illustrates the basic conditions that must be in place in order to observe a rainbow. With the sun at the viewers back and rain falling at some distance in front, the sunlight that is incident on the falling raindrops (essentially solid spheres of water) is refracted, dispersed, and reflected back to the viewer. The geometry of the situation results in the red light being bent through an angle that is about 2° less than the blue light. The result is that we see a semicircular band of light across the sky. The width of that band subtends a 2° angle, with its color ranging from red on the outside, yellow and green at its center, to blue on its inside. Under extraordinary conditions it is sometimes possible to observe a second, considerably less bright, rainbow inside the primary. The spectral order of this secondary rainbow will be reversed relative to that of the primary, because it has been created by a second reflection within the raindrop.

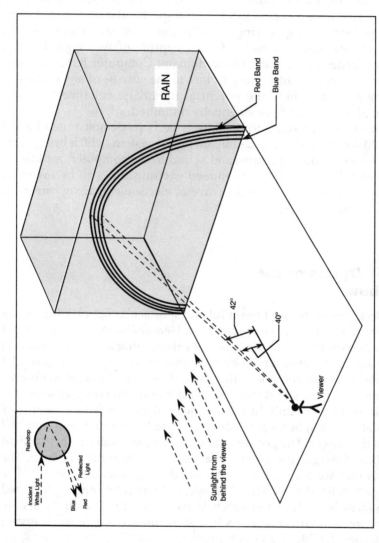

Figure 12-8. The phenomenon of the rainbow will be observed when conditions such as those shown here are in place. The spherical raindrop will act as a combination lens and mirror to disperse the incident sunlight and reflect it to the viewer's eye.

12.10 Review and Summary

This chapter has been included in the hope that it will convey to the reader some measure of the fun and excitement that can be a part of optics and optical engineering. The discussion of optical illusions illustrates the complexity and sophistication of the human visual system, which includes our built-in camera (the eye), our detector array (the retina), and a data-processing capability beyond that of most computers. A brief discussion of three-dimensional, or stereo, viewing has also been included, with reference to applications to the field of entertainment along with more serious realms.

Modern technology abounds with developments that are centered around the science of optics and the work of the optical engineer. Home entertainment has been revolutionized by digital recording and optical playback schemes applicable to both sound and video systems. Professional and amateur photography have benefited greatly from recent advances in lens design, camera system design, and new, better films. Developments in the field of electronic imaging appear to be leading inexorably toward a time when film, as we know it, will become obsolete and the entire field of image recording and viewing will be dominated by electronics, computers, and microprocessors.

A basic understanding and appreciation of the fundamental principles as presented throughout this book should serve you well in preparing you to observe and appreciate many of the optical phenomenon that you may encounter on a daily basis. And, in the best possible scenario, you will be inspired to explore further this most exciting and rewarding field, perhaps choosing it for your ultimate life's work.

Appendix $\mathbf{A}$

Basic Optical Engineering Library

This addendum contains a listing and description of available textbooks and reference literature that persons working in the field of optical design will find informative and helpful. While deserving of considerable and justifiable debate, the order in which the "top 10" have been presented reflects this author's opinion regarding their relative importance and usefulness, based on my experience while working in the field.

1. Warren J. Smith, *Modern Optical Engineering*, McGraw-Hill, 1990. Since it was first published in 1962, *Modern Optical Engineering* has become the standard text in the field of optical engineering. It presents a unique combination of the required traditional concepts, that have been taught for decades, with the modern ideas and methods employed widely throughout the optics industry today.

2. Rudolph Kingslake, *Applied Optics and Optical Engineering*, Academic Press, 1965. This is a series of books dealing with optical theory, design, and optical engineering. The first five books in the series are especially useful to the engineer in need of general tutorial and reference information. Subsequent books in the series are dedicated to more specific topics, and they are highly recommended to persons working in those areas.

3. David Falk, Dieter Brill, and David Stork, *Seeing the Light*, Wiley, 1986. This is a very interesting and informative book dealing with the subjects of physics and light as they are encountered in nature and in science. It covers such topics as geometric optics, photography, vision, physical (wave) optics, and holography. The information on the human visual system, including the perception of color, is particularly well done.

4. Robert E. Hopkins and Richard Hanau, *Military Handbook 141*, republished by Sinclair Optics, 1987. First published by the government in the early 1960s, *MIL Handbook 141* has served as a basic reference to optical engineers and designers since that time. While no longer available from the government, Sinclair Optics has republished the key sections dealing with geometric optics and optical (lens) design. The handbook contains a complete treatment of the basic methods employed in the design of optical instruments.

5. *The Photonics Directory*, Laurin Publishing Co. (annual). *The Photonics Directory* consists of four separate volumes: *The Corporate Guide, The Buyers' Guide, The Photonics Handbook,* and *The Photonics Dictionary*. The corporate and buyers' guides offer a comprehensive picture of the photonics industry, on an international scale. The handbook and dictionary provide the reader with a wealth of valuable and useful reference information.

6. Rudolph Kingslake, *A History of the Photographic Lens*, Academic Press, 1989. This book conveys to the reader a sense of the history of the optics industry, and especially the history of lens design as it relates to the field of photography. A recognized authority in this field, Dr. Kingslake has contributed significantly to the training of many of today's leading optical engineers while at the University of Rochester.

7. Warren J.Smith, *Modern Lens Design*, McGraw-Hill, 1992. Primarily a resource manual, *Modern Lens Design* contains some very valuable information dealing with automatic lens design, optimization, and evaluation. This is followed by a comprehensive presentation of complete lens design and aberration data on essentially all lens types. These designs will often serve well as starting points in the generation of new lens designs.

8. Milton Laikin, *Lens Design*, Marcel Dekker, 1991. Similar to *Modern Lens Design*, this book presents a section on lens design methods, followed by complete prescription data on a number of interesting and useful lens designs, with MTF data. Several appendices contain valuable reference information.

9. Francis W. Sears, *Optics*, Addison-Wesley, 1949. This is a basic text which deals with the fundamental concepts of both physical and geometric optics. Prepared for use in the teaching of general physics, this book serves well as a basic reference. Its vintage precludes any coverage of modern lens design techniques.

10. Jurgen R. Meyer-Arendt, *Introduction to Classical and Modern Optics*, Prentice-Hall, 1972. This is a very interesting book which deals primarily (and quite nicely) with the subject of physical optics and electromagnetic wave theory. Especially valuable and interesting is the complete and thorough treatment of the history of optics and the key persons involved with the science over the years.

11. Francis A. Jenkins and Harvey E. White, *Fundamentals of Optics*, McGraw-Hill, 1957. This is a sweeping treatment of the subject. A section on geometric optics deals with basic components and optical instruments. More time is spent on the subject of physical optics, wave theory, and interference. Finally, a portion of the book presents a discussion of quantum optics and photon theory.

12. Donald H. Jacobs, *Fundamentals of Optical Engineering*, McGraw-Hill, 1943. This book deals with geometric optics and instrument design. Based on activity during World War II, the presentation is refreshingly basic, and it concentrates on fundamental principles. Fortunately for the reader, much of the information contained is timeless, while the presentation is very clear and easy to follow.

13. Rudolph Kingslake, *Lenses in Photography*, A. S. Barnes & Co., 1963. This book will be especially useful to the optical engineer involved in the field of photography. Much valuable and interesting historic information is presented, along with considerable technical data and reference information difficult to locate elsewhere.

14. Earle B. Brown, *Modern Optics*, Reinhold, 1965. *Modern Optics* was written at a time when the optical industry was in a state of general turmoil and rapid growth. As a result, it reflects the status at a most interesting time. General system and instrument design is covered, along with an interesting introduction to the laser and its applications.

15. Philip Kissam, *Optical Tooling*, McGraw-Hill, 1962. This book deals with the general subject of optical instrumentation and optical tooling in particular. The book was written at a point in time when the subject was peaking in its popularity and applications. It is well written and contains considerable useful information regarding optical tooling instruments and accessories of that time.

16. John Strong, *Concepts of Classical Optics*, Freeman, 1958.   This textbook was written to be used in teaching an intermediate optics course. It is very well written and beautifully illustrated. It covers both physical and geometric optics, with much information relating to the manufacture and test of precision optics.

17. Max Born and Emil Wolf, *Principles of Optics*, Pergamon Press, 1965.   This is the ultimate text on the subject of electromagnetic theory, interference, and diffraction. The mathematical presentations are rigorous and will be challenging to most readers. This is the definitive book dealing with physical optics.

Appendix B
Optical Design Software Sources

This addendum contains a listing of the principal sources of optical design software at the date of publication (1994). The vendors appearing here offer software packages that are capable of complete optical design, optimization, and image evaluation. The dynamic nature of this field suggests that the reader consult the most recent reference and trade publications available to obtain up-to-date information relative to these companies, as well as possible newcomers to the field.

Breault Research Organization, Inc.
4601 E. First Street
Tucson, AZ 85711
(602) 795-7885

Focusoft, Inc.
P.O. Box 756
Pleasanton, CA 94566
(510) 426-1835

Kidger Optics Ltd.
9A High Street
Crowborough, East Sussex TN6 2QA
England
0892 663555

Optical Research Associates
3280 E. Foothill Blvd.
Pasadena, CA 91107
(818) 795-9101

Optikos Corporation
286 Cardinal Medeiros Ave.
Cambridge, MA
(617) 354-7557

Sciopt Enterprises
P.O. Box 20637
San Jose, CA 95160
(408) 268-9050

Sinclair Optics, Inc.
6780 Palmyra Road
Fairport, NY 14450
(716) 425-4380

Optical Glass Sources

This addendum contains a listing of the principal sources of optical glass at the date of publication (1994). The vendors appearing here have been separated into two categories: primary and secondary. Primary sources manufacture glass from raw materials and supply blanks in a variety of forms. Secondary sources produce specialty materials or procure glass stock from primary suppliers and modify the form to suit the customers' needs. The unusual nature of this field suggests that the reader consult the most recent reference and trade publications available to obtain up to date information relative to these companies, as well as possible newcomers to the field.

Primary Sources

Schott Glass Technologies, Inc.
400 York Avenue
Duryea, PA 18642-2026
(717) 457-7485

Ohara Corporation
50 Columbia Road
Branchburg
Somerville, NJ 08876-3518
(908) 218-0100

Hoya Optics, Inc.
3400 Edison Way
Fremont, CA 94538-6190
(510) 490-1880

Sumita Optical Glass, Inc.
P.O. Box 37
Briarcliff Manor, NY 10510
(914) 762-2639

Secondary Sources

Corning Incorporated
Building 21-4-2
Corning, NY 14831-0001
(510) 426-1835

Glass Fab, Inc.
P.O. Box 1880
Rochester, NY 14603
(716) 262-4000

Libbey-Owens-Ford Co.
811 Madison Avenue
Toledo, OH 43695-0799
(419) 247-3931

Newport Glass Works, Ltd.
1629 Monrovia Avenue
Costa Mesa, CA 92627
(714) 642-9980

United Lens Co., Inc.
259 Worcester Street
Southbridge, MA 01550-1325
(508) 765-5421

Conversion Factors and Constants

Acceleration due to gravity	$g = 980.67 \text{ cm/s}^2 = 32.174 \text{ ft/s}^2$
Ampere	$1 \text{ A} = 1 \text{ coulomb (C)/s}$
Angstrom	$1 \text{ Å} = 1 \times 10^{-8} \text{ cm} = 1 \times 10^{-1} \text{ μm}$
Atmosphere	$1 \text{ atm} = 760 \text{ mm Hg} = 14.696 \text{ lb/in}^2$ $= 1.013 \times 10^6 \text{ dyn/cm}^2 = 1033$ g/cm^2
Atomic mass unit	$1 \text{ amu} = 931 \text{ MeV}$
Avogadro's number	$N = 6.022 \times 10^{23}$
Boltzmann's constant	$k = 1.3807 \times 10^{-16} \text{ ergs/K}$
British thermal unit	$1 \text{ Btu} = 252 \text{ calories}$
Calorie	$1 \text{ cal} = 4.184 \text{ joules (J)} = 0.04129$ liter atm
Centimeter	$1 \text{ cm} = 0.01 \text{ m} = 0.3937 \text{ in}$
Centimeter/second	$1 \text{ cm/s} = 0.02237 \text{ mi/h}$
Cubic centimeter	$1 \text{ cm}^3 = 0.06102 \text{ in}^3$
Cubic inch	$1 \text{ in}^3 = 16.387 \text{ cm}^3$
Density	$D_{\text{water}} = 1.000 \text{ g/cm}^3 \text{ at } 4°\text{C}$ $D_{\text{mercury}} = 13.6 \text{ g/cm}^3 \text{ at } 0°\text{C}$ $D_{\text{air}} = 1.293 \times 10^{-3} \text{ g/cm}^3 \text{ (at STP)}$
e (natural log base)	$e = 2.7183$

Electronic charge	$e^- = 4.80 \times 10^{-10}$ esu $= 1.60 \times 10^{-19}$ C
Electronvolts per atom	1 eV/atom = 23.05 kcal/mol
Erg	1 erg $= 2.389 \times 10^{-8}$ cal $= 1 \times 10^{-7}$ J
Faraday	1 Faraday = 96,500 C = 23,070 cal/V $= 6.023 \times 10^{23} e^-$
Gas constant	$R = 0.08205$ liter atm/mol °K $= 1.987$ cal/(mol °K) $= 8.314 \times 10^7$ ergs/(mol °K)
Gram	1 g mass $= 2.15 \times 10^{10}$ kcal
Grams per cubic centimeter	1 g/cm^3 = 62.43 lb/ft^3
Gram molecular volume	$V_0 = 22.413$ liter/mol at 0°C, 1 atm
Ice point	$T_0 = 273.15$ K
Inch	1 in = 2.540 cm
Joule	1 J = 0.2390 cal
Kilogram	1 kg = 2.205 lb
Liter	1 liter = 1000 cm^3 = 1.0567 quarts (qt)
Liter-atmosphere	1 liter atm = 24.22 cal
Natural logarithms	$\ln x = 2.3026 \log_{10} x$
Mass of earth	$M = 5.983 \times 10^{24}$ kg
Pi	$\pi = 3.1416$
Planck's constant	$h = 6.626 \times 10^{-27}$ erg s
Pound	1 lb = 453.6 g
Quart	1 qt = 0.9463 liter
Speed of light	$c = 2.998 \times 10^{10}$ cm/s
Speed of sound	$v = 331.7$ m/s (in air at STP) = 1470 m/sec (in water at 20°C)
Standard temperature and pressure	STP = 0°C/760 mm Hg
Stefan-Boltzmann constant	$\sigma = 5.6705 \times 10^{-5}$ erg/cm$^2 \cdot$ s (K)^4

Measures and Equivalents

Length	Equivalent
1 millimeter	0.001 meter
	0.03937 inch
1 nanometer	1×10^{-9} meter
1 centimeter	10 millimeters
	0.3937 inch
1 inch	25.4 millimeters
1 decimeter	10 centimeters
	3.937 inches
1 foot	30.48 centimeters
	12 inches
1 meter	100 centimeters
	39.37 inches
1 yard	0.9144 meter
	3 feet
1 fathom	6 feet
1 dekameter	10 meters
	1.9884 rods
1 rod	0.5029 dekameter
	5.5 yards

1 furlong	40 rods
	0.125 miles
1 hectometer	100 meters
1 kilometer	1000 meters
	0.6214 miles
1 statute mile	1609.3 meters
	5280 feet
	8 furlongs
1 nautical mile	6076 feet
	1.15 statute miles
1 league	3 nautical miles
1 light year	9.4637×10^9 meters
	5.8804×10^{12} miles

Area	**Equivalent**
1 square centimeter	100 square millimeters
	0.1550 square inch
1 square inch	6.4516 square centimeters
1 square decimeter	100 square centimeters
	0.1076 square foot
1 square foot	9.2903 square decimeters
	144 square inches
1 square meter	100 square decimeters
	1.196 square yards
1 square yard	0.8361 square meter
	9 square feet
1 square dekameter	100 square meters
1 square hectometer	100 square dekameters
1 square kilometer	100 square hectometers
	0.3861 square mile
1 square mile	640 acres
	2.59×10^6 square meters
1 acre	160 square rods
	43,560 square feet
	4046.9 square meters

Volume and Capacity / Equivalent

Volume and Capacity	Equivalent
1 cubic centimeter	0.06102 cubic inch
1 cubic inch	0.0164 liter
1 cubic decimeter	0.0353 cubic foot
1 cubic foot	28.32 liters
1 cubic meter	1.308 cubic yards
1 liter	1,000 cubic centimeters
	61.024 cubic inches
	0.0353 cubic foot
	0.9081 quart dry
	0.0284 bushel
	1.0567 quarts liquid
	2.2046 pounds (lb) of water (4°C)
1 dekaliter	2.6417 gallons

Time / Equivalent

Time	Equivalent
1 day	1440 minutes
	86,400 seconds
1 week	7 days
1 sidereal year	365.256 days
	8766.144 hours

Velocity / Equivalent

Velocity	Equivalent
1 meter per second (m/s)	2.2369 miles per hour
1 mile per hour (mi/h)	1.4667 feet per second (ft/s)
	1.6093 kilometers per hour (km/h)
	0.447 meter per second (m/s)
1 knot	1 nautical mile per hour
	1.1508 statute miles per hour
1 radian per second	9.549 revolutions per minute
1 revolution per minute (rpm)	0.10472 radian per second

Angle / Equivalent

Angle	Equivalent
1 radian	$360/2\pi$ degrees
	57.296 degrees

1 degree	0.01745 radian
	60 minutes
1 minute	2.909×10^{-4} radian
	60 seconds
1 second	4.85×10^{-6} radian
1 miliradian	0.001 radian
	0.0573 degree
1 mil (military)	360/6400 degree (°)
	0.05625 degree
	0.00098 radian
1 solid angle	4π steradians
	1 sphere
1 steradian	0.07958 solid angle

Weight	**Equivalent**
1 gram	0.03527 ounce
	15.4354 grains
1 kilogram	2.2046 pounds
1 quintal	100 kilograms
1 ton (English)	10 quintals
1 ton (metric)	1.1023 tons (English)
1 short ton (Avoirdupois)	2000 pounds
1 long ton (Avoirdupois)	2240 pounds

Energy and Power	**Equivalent**
1 horsepower	746 watts (W)
	33,000 ft lb/min
	2,544 Btu/h
1 horsepower-hour	273,740 kg meters
1 foot-pound	1356 J
	0.13826 kg meters
1 watt	1 J/s
	3.413 Btu/h
	44.22 ft lb/min
1 kilowatt	1.34 horsepower
	56.9 Btu/min

1 British thermal unit (Btu)	1055 W s
	778 ft lb
1 joule	0.73756 ft lb

Acceleration — Equivalent

| 1 ft/s^2 | 30.480 cm/s^2 |
| | 0.6818 mi/h · s |

Pressure — Equivalent

1 kg/cm^2	14.223 lb/in^2
	0.9678 normal atmosphere
1 kg/m^2	0.2048 lb/ft^2
1 lb/in^2	6894.7 pascals
	27.71 in of water
	2.04 in of mercury
	0.06805 normal atmosphere
1 lb/ft^2	4.882 kg/m^2
1 in of water	0.0361 lb/in^2
	0.0735 in of mercury
1 in of mercury	0.4912 lb/in^2
	13.58 in of water
1 normal atmosphere	1.0133 bars
	1.0332 kg/cm^2
	14.696 lb/in^2
	33.95 ft of water
	760 mm of mercury
1 pascal (Pa)	0.000145 lb/in^2

Force — Equivalent

1 newton	1×10^5 dyn
	0.22481 lb
1 dyne (dyn)	2.2481×10^{-6} lb
	7.233×10^{-5} poundal
1 poundal	0.03108 lb
	1.3825×10^4 dyn

Torque

1 dyn · cm

Equivalent

1.0197×10^{-8} kg · m

7.3757×10^{-8} ft lb

2.3731×10^{-6} ft poundal

Temperature

Δ 1 degree Centigrade (°C)

0°C

100°C

Equivalent

Δ 1.8 degrees Fahrenheit (°F)

32°F

212°F

Basic Photometric Considerations

The subject of photometry deals with the quantities of visible light that occur at the scene being viewed or recorded, and the corresponding light level at the system detector. This appendix contains information that will be helpful in understanding a few of the fundamental points involved. First is a brief review of the *exposure value* (E_v) system.

The Basic Factors

Correct exposure results when the amount of light delivered to the image plane is in agreement with the light level required by the sensor to produce an optimum final image. There are just four contributing factors that must be considered in order to ensure that this balance has been achieved:

- The brightness of the scene
- The sensitivity of the film
- The f number of the optics
- The time (duration) of the exposure

The relationship between factors is demonstrated by the fact that the effect of either a brighter scene, or a more sensitive film, can be balanced by a shorter exposure time, or a smaller aperture setting (larger f number). A more useful and detailed understanding of these relationships will result from a review of the exposure value, or E_v system.

The E_v System

The E_v system was developed in the 1950s by German and American groups involved in scientific standardization. The basic information generated at that time has been updated, expanded, and is presented in tabular format at the end of this appendix (Fig. F-1).

Figure F-1 shows how each of the exposure factors, system f number (A_v), shutter speed (T_v), sensor ISO rating (S_v), and scene brightness (B_v), can be assigned a value ranging from -3 to $+11$. Each incremental increase in that value represents a doubling of the factor's absolute value. The key to applying the information in Fig. F-1 is contained in the basic exposure value formula:

$$E_v = A_v + T_v = S_v + B_v$$

From this formula we see that, for correct exposure, the sum of the system f number and shutter speed values, must be equal to the sum of the sensor ISO rating and scene brightness values.

Using the Table

The use of the E_v system and the tabulated information is best demonstrated by applying it to two typical examples. For the first, assume that we are about to embark on a photographic excursion on a typical fall day. Since the camera will be handheld, a shutter speed of at least $\frac{1}{60}$th second ($T_v = 6$) will be required. For best color rendition we select a fine-grain color slide film with an ASA rating of 25 ($S_v = 3$). As the sun passes in and out of clouds, the scene brightness will vary from bright overcast to bright sunny ($B_v = 8 \pm 2$). Plugging this data into the E_v equation, we have

$$E_v = A_v + T_v = S_v + B_v$$
$$A_v + 6 = 3 + 8\,(\pm 2)$$
$$A_v = 5\,(\pm 2)$$

From Fig. F-1 we can see that this value for A_v corresponds to lens aperture setting of $f/5.6$ on average, with a maximum of $f/11$ and a minimum of $f/2.8$, depending on the scene brightness. If it is felt that at $f/2.8$ the image quality, or depth of field, might not be satisfactory, we could improve on that situation by choosing a more sensitive film. Selection of a film with an ASA rating of 100 ($S_v = 5$), for example, will increase the E_v value by 2, resulting in a more desirable lens aperture range of from $f/5.6$ to 22.

A_V	f/#	T_V	Shutter Speed (Seconds)	S_V	ISO Rating	B_V	Scene Brightness (Ft L)	Scene Illuminance (Lux)	Scene Illuminance (Ft C)	Typical Exterior	Typical Interior
11	44	11	1/2000	11	6400	11	2000	120 K	11 K		
10	32	10	1/1000	10	3200	10	1000	60 K	5600	Sunny Day	
9	22	9	1/500	9	1600	9	512	30 K	2800		
8	16	8	1/250	8	800	8	256	15 K	1400	Open Shade	
7	11	7	1/125	7	400	7	128	8 K	740		
6	8	6	1/60	6	200	6	64	4 K	360		
5	5.6	5	1/30	5	100	5	32	2 K	180	Overcast Day	Well Lit Arena
4	4	4	1/16	4	50	4	16	1 K	90		
3	2.8	3	1/8	3	25	3	8	480	45		
2	2	2	1/4	2	12	2	4	240	22		
1	1.4	1	1/2	1	6	1	2	120	11	Sunrise Sunset	Daytime
0	1	0	1	0	3	0	1	60	5.6		
-1	.7	-1	2	-1	--	-1	.5	30	2.8	Twilight	
-2	.5	-2	4	-2	--	-2	.25	15	1.4		Nighttime
-3	--	-3	8	-3	--	-3	.12	.7	0.6		

Exposure Value = E_V = A_V + T_V = S_V + B_V

Figure F-1. The exposure value (E_v) system.

For a second example, assume we wish to take photos at a sporting event in a well-lit indoor arena where the scene brightness will average about 16 foot-lamberts (ft · L) ($B_v = 4$). Selection of a zoom lens limits us to a maximum lens aperture of $f/4$ ($A_v = 4$). The fast-paced action would dictate a shutter speed of about $\frac{1}{250}$th second ($T_v = 8$). This allows us to plug three of the four factors into the E_v equation and solve for the fourth as follows:

$$A_v + T_v = S_v + B_v$$
$$4 + 8 = S_v + 4$$
$$S_v = 8$$

From the table, we find that an S_v value of 8 corresponds to a rather fast film with an ISO rating of 800.

An Alternate Application

A basic understanding of the E_v system, in conjunction with the information given in Fig. F-1, makes it possible to utilize a handheld light meter, or a camera with built in meter, to determine scene brightness or illuminance in absolute terms.

For example, suppose we wish to determine the light level that exists at our desktop. With a handheld light meter, set to an ASA setting of 100 ($S_v = 5$), a meter reading of reflected light can be taken by pointing the meter at the desktop. Assume that the meter indicates the scene should be photographed using a lens aperture of $f/4$ ($A_v = 4$), using a shutter speed of $\frac{1}{30}$th of a second ($T_v = 5$).

Now, the scene (desktop) brightness (B_v) can be found:

$$B_v = A_v + T_v - S_v = 4 + 5 - 5 = 4$$

From Fig. F-1 we find that a B_v of 4 corresponds to a brightness of 16 ft-lamberts and (assuming a standard scene reflectance of 18 percent) a desktop illuminance of 960 lux, or 90 footcandles (fc).

Summary of the E_v System

Use of the standard exposure value formula and the data in Fig. F-1 leads to a better understanding of those factors involved in determining correct exposure. With this as a guide, it is possible to evaluate most imaging situations in advance and determine the best approach to achieving properly exposed images. While this method will yield reasonable estimates, best results will be achieved through the use of a modern light metering system to confirm these estimates and to make the ultimate determination of the correct exposure setting.

In an alternate application, we have seen that it is possible to dust off our rarely used light meter, or use our camera's built-in meter, in conjunction with the data presented here, to determine absolute light conditions at a nearby or remote location.

Image Illuminance

Modern systems dealing with electronic imaging will frequently specify the performance of the system detector as a function of the illumination level that is present at that detector. The formula given in Fig. F-2 shows the relationship that exists between illuminance at the scene, and illuminance at the image (detector) plane. Applying this formula, it is possible to determine the range of lens speeds that must be provided by the system in order to image a designated range of scene brightness with a given detector. When it is found that the range of scene brightness is greater than can be accommodated by the system f number, it is possible to modify the system transmission by the incorporation of switching, or variable, neutral density filters.

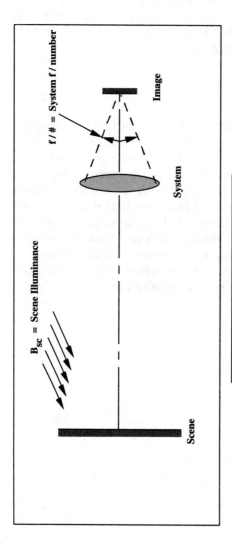

Figure F-2. Image illuminance.

$$B_{im} = \frac{B_{sc} \cdot R \cdot t}{4 \cdot f/\#^2}$$

where:

B_{im} = Image Illuminance

B_{sc} = Scene Illuminance

R = Scene Reflectance

t = System transmission

$f/\#$ = System f / number

Surface Sag and Conic Sections

Surface Sag (Sphere)

During the process of optical system design and layout, it is often necessary to determine the depth, or sag, of an optical surface at some specific height. For example, consider the configuration of a simple lens cell, as shown in Fig. G-1a. In order to protect the glass from physical damage, it is necessary to determine the sag of the convex lens surface at the height designated Y. The exact formula for the surface sag S is

$$S = R - (R^2 - Y^2)^{1/2}$$

When the ratio of R to Y is greater than 5:1, the following formula can be used to calculate the approximate sag, with an error of less than 1 percent

$$S \approx Y^2/2R$$

The Parabola

By far the most common shape for an optical surface is spherical. The spherical surface is created by rotating a circular section about the optical axis. Another surface shape that is frequently encountered in

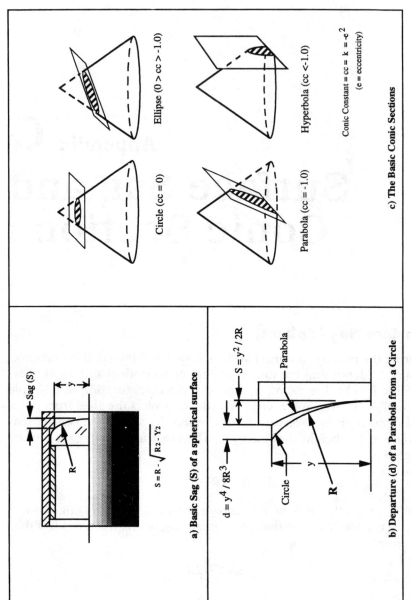

Figure G-1. Surface sag and conic sections.

The following labels appear within the figure:

a) Basic Sag (S) of a spherical surface

Sag (S)

y

R

$$S = R - \sqrt{R^2 - Y^2}$$

b) Departure (d) of a Parabola from a Circle

$S = y^2 / 2R$

Parabola

$d = y^4 / 8R^3$

Circle

y

R

c) The Basic Conic Sections

Ellipse (0 > cc > -1.0)

Circle (cc = 0)

Hyperbola (cc < -1.0)

Parabola (cc = -1.0)

Conic Constant = cc = $k = -e^2$

(e = eccentricity)

optical system design is the paraboloid. In this case a parabolic cross-section is rotated about the optical axis to produce the solid paraboloidal surface. The paraboloid has the unique—and frequently very useful—characteristic of focusing all light rays in a collimated on-axis light bundle to a common point when the paraboloid is in the form of a concave reflector. In other words, the parabola is completely free of on-axis aberrations.

In the case of the paraboloid, the formula given above for the approximate sag of a sphere, becomes exact. For a parabolic surface:

$$S = \frac{Y^2}{2R}$$

Figure G-1b shows the relationship between the parabola and the circle, regarding their respective sag formulas and the departure of the parabola from its reference (vertex) circle.

Other Conic Sections and the Conic Constant

Two other surface shapes are derived from cross sections taken through a solid right-angle cone: the ellipse and the hyperbola. Any of these conic sections can be described for purposes of optical analysis by designating their vertex curvature and conic constant.

Figure G-1c shows a plane intersecting a cone in four different orientations. In the upper left, the plane is parallel to the base of the cone and the resulting cross section is a circle. In the upper right, the plane is tipped downward and the resulting cross section is an ellipse. When the plane is tipped to the point where it becomes parallel with the far side of the cone (lower left), the cross section that results is a parabola. Finally, when the plane is tipped such that it is perpendicular to the base of the cone (lower right), the cross section that results is a hyperbola.

In most optical design software, the conic surfaces are described by designating their vertex radius along with a conic constant (cc or k). Figure G-1c indicates the corresponding conic constant for each of the four common conic sections.

Index

About the Author

Bruce H. Walker, founder and president of Walker Associates, has been active in the fields of optical engineering and lens design since 1960. His initial work was with General Electric, where he received four patents for unusual lens designs. He was with the electro-optical division of Kollmorgen Corp. for 20 years, first as a senior optical engineer and then engineering manager. Since 1970, Mr. Walker has been a member of the Editorial Advisory Board of The Laurin Publishing Company. During that time he has had over 30 articles published and has contributed significantly to their *Photonics Handbook and Dictionary*. Since 1990, he has worked as an independent consultant, specializing in the solution of optical engineering problems and the generation of specialized lens designs.